Risk Propagation Assessment for Network Security

FOCUS SERIES IN NETWORKS AND TELECOMMUNICATIONS

Series Editor Marcelo Dias de Amorim

Risk Propagation Assessment for Network Security

Application to Airport Communication Network Design

Mohamed Slim Ben Mahmoud
Nicolas Larrieu
Alain Pirovano

WILEY

First published 2013 in Great Britain and the United States by ISTE Ltd and John Wiley & Sons, Inc.

ISTE Ltd
27-37 St George's Road
London SW19 4EU
UK

www.iste.co.uk

John Wiley & Sons, Inc.
111 River Street
Hoboken, NJ 07030
USA

www.wiley.com

Library of Congress Control Number: 2012954206

British Library Cataloguing-in-Publication Data
A CIP record for this book is available from the British Library
ISSN: 2051-2481 (Print)
ISSN: 2051-249X (Online)
ISBN: 978-1-84821-454-5

Printed and bound in Great Britain by CPI Group (UK) Ltd., Croydon, Surrey CR0 4YY

Contents

List of Figures

List of Tables

Introduction

Risk assessment is generally considered as the core of the computational framework in a risk management process for a network information system. This process is mandatory and crucial for the protection of interconnected systems that provide various services to their clients or users. Traditionally, system vulnerabilities are identified, determining the occurrence of threats being exploited, and evaluating the consequences of attacks that could take advantage of these security weaknesses. Having assessed the risks, security countermeasures (which could be technical or operational) are identified and then implemented to mitigate those risks.

In the literature, there is a plethora of risk assessment methodologies defined for network security. Nevertheless, most of them rely on a quantitative security analysis. In such approaches, security experts evaluate the potential likelihood and impact of each identified threat to determine the risk level for the network system. This is a big issue as security expertise is expensive financially speaking, and relatively slow compared to an automated risk assessment procedure: several workshops and appointments have to be held in order to catch the subjective assessment of the experts. Furthermore, the assessment risk processes are often based on a ranking scale (e.g. low, medium, high). Using such an evaluation methodology allows us possibly to compare two different risk levels (for instance, between high and low), but it is impossible to estimate the distance between them (for instance, between two security levels ranked as high). This can be confusing for a security network administrator willing to improve the overall security level of the network.

The *Improving Network Security Using a Risk Propagation-based Quantitative Assessment Methodology* book deals with risk assessment methodologies for network architecture design. The main goal is to present and illustrate an innovative risk propagation-based quantitative assessment tool. This risk assessment methodology takes into account the inherent characteristics of any computer network in general (such as interconnection between nodes). This contribution is motivated, first by the shortcomings noticed in qualitative risk assessment methodologies. Second, the existing quantitative risk assessment models for network security generally missed out an essential concept in network security risk evaluation, namely risk propagation. In this book, we fill this gap showing how important risk propagation can be in the network risk computation step. Furthermore, this original approach helps network and security administrators to design and build more robust and secure network topologies.

As an implementation case study, a new aeronautical communication system called AeroMACS (Aeronautical Mobile Airport Communications System) is considered. AeroMACS has been recently identified as the future wireless access network for airport surface communications, which will be soon deployed at European and American airports. It is based on the IEEE 802.16-2009 standard, which is also known as WiMAX[1]. Indeed, aeronautical communications are about to shift the paradigm of digital data in the near future.

Owing to the progress made in **information technology (IT)** in the last decades, aviation industry stakeholders are enhancing and

1 http://standards.ieee.org/about/get/802/802.16.html

expanding their networks not only to overcome congestion issues related to voice radio communications but also to modernize the **air traffic management (ATM)** environment through worldwide collaborative projects like Single European Sky ATM Research **(SESAR)**. The introduction of data communication technologies, such as AeroMACS, emphasizes the need for secure systems to protect against cyber attacks and external threats.

This book is split into two parts. Part 1 is related to network security risk assessment, while Part 2 concerns the aeronautical case study and the AeroMACS technology.

Part 1: Network security risk assessment

Chapter 1 provides an introduction to information system security risk management process. The objective of this chapter is to underline the importance of network security in network design and emphasize the impact of risk assessment in the decision-making process of network security policies. Furthermore, the risk propagation concept is explained.

Chapter 2 provides a state of the art of existing risk management methodologies that can be currently used (quantitative and qualitative **methodologies**).

In Chapter 3, an original quantitative network risk assessment model based on risk propagation is introduced. All the parameters defined in the methodology are explained. Then, the chapter goes through the network security risk assessment process.

Part 2: Application to airport communication network design

After the first part dedicated to risk assessment, an extensive case study is presented in the second part of the book. The main goal is to present to the reader a complete application of the risk assessment methodology to future airport communication network design.

Chapter 4 introduces the aeronautical background necessary to understand this case study: the European SESAR project is briefly presented and then specific aeronautical communication concepts and terminology are discussed. The chapter ends with a section dedicated to the AeroMACS technology (protocol architecture, network reference model, security features) used in the case study.

Finally, Chapter 5 presents all the experimental scenarios and their results. The risk assessment results are used at each step to improve the choices made later in the following experiments as the tool is meant to be used (a decision-making tool for network security and network design).

PART 1

Network Security Risk Assessment

1

Introduction to Information System Security Risk Management Process

Currently, network security is an important part of a network design process. Information System Security Risk Management (ISSRM) allows network engineers to maximize the network security level they want to achieve. Usually, ISSRM processes follow an overall framework composed of classical and common steps. Nevertheless, these steps can differ from one method to another and do not necessarily put the same weight on each step. For instance, some methods focus on security controls and countermeasures whereas others put more effort on risk assessment and treatment procedures.

However, a general ISSRM framework can be drawn and considered as a basis for any information security management-related work, as illustrated in Figure 1.1.

This general ISSRM framework is composed of six steps:

a) *Context and asset identification*: first, the communication system and its environment are described with an emphasis on the sensitive assets (devices, data, etc.) to protect.

b) *Security objectives identification*: security needs are then defined. Based on the previous step, security objectives are usually expressed in terms of basic security services (confidentiality, authentication, integrity, non-repudiation, availability and authorization).

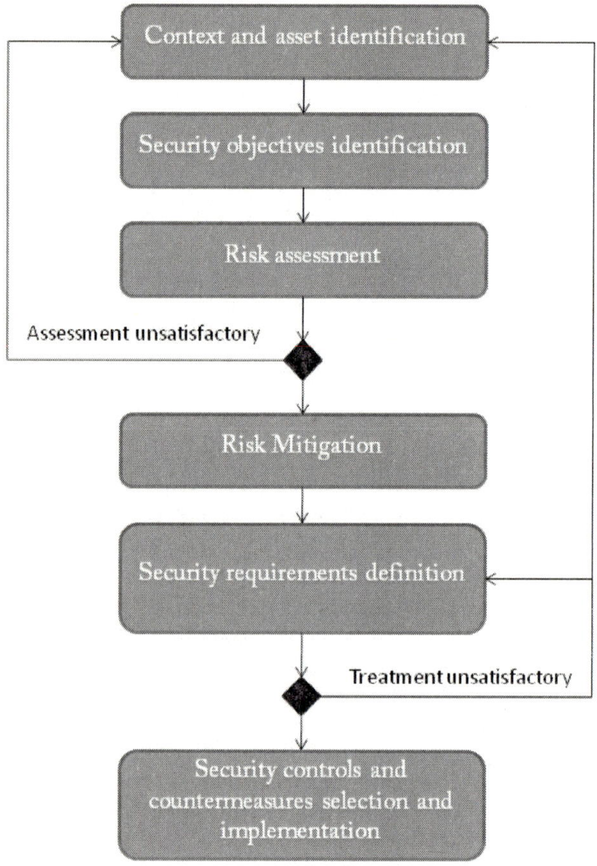

Figure 1.1. *General information system security risk management process*

c) *Risk assessment*: this step consists in estimating potential risks that can harm the assets identified in step (a) and threaten the security objectives of step (b). The risk assessment procedure can be based on a qualitative or quantitative study. Note that if the risk assessment is unsatisfactory, it could be possible to go back to previous steps and restart the analysis.

d) *Risk mitigation*: once the risk has been clearly identified, risk treatment actions can be taken. For instance, such a measure could be to decide to retain the risk (e.g. accept the risk because it is considered low enough), reduce it (e.g. reinforce security policies) or avoid it (e.g. deactivate a risky network device).

e) *Security requirements definition*: security requirements can now be determined as security solutions to mitigate the identified risks, mainly if the risk reduction strategy has been chosen.

f) *Security controls and countermeasures selection and implementation*: finally, security requirements are instantiated into explicit security controls and countermeasures. For instance, firewalls have been selected to protect the aeronautical network we consider in the case study developed in Chapter 5 of this book.

1.1. On the importance of network security for network designers

Network security is a critical step in IT network and system design. Security countermeasures (firewalls, virtual private networks (**VPNs**), authentication, authorization and accounting (**AAA**) servers, etc.) are the first protection layer against threats and malicious actions targeting the system resources. In order to provide an effective and robust network, a sound risk analysis and a well-thought security policy are required. Hence, before deploying the network security system, network designers have to carefully think about security by respecting the following guidances:

– Security has to be a *built-in* feature instead of a *built-on* ingredient to be added when security issues begin to show up.

– Several layers of security should be deployed in order to complement each other when needed (e.g. if a firewall goes offline, another firewall will be able to provide data flow filtering). Also, security devices should be physically located into different entries of the network in order to avoid a single “point of failure”. This is usually called “*defense-in-depth*” security.

– IT system resources should be dispatched into different security demilitarized zones (**DMZs**) according to their sensitivity and criticality.

– Intrusion prevention systems (**IPS**) are different from firewalls and should be used because they do not perform the same role: IPS devices are about intrusion detection for later effective actions whereas firewalls are about monitoring traffic flows, compartmentalizing the network infrastructure, and controlling the communications.

– Network security should not be limited to security devices only, it should be extended to other network management and configuration tasks (e.g. setting configuration files on network devices, monitoring resource consumption activities, analyzing logging events).

– A good network security design necessary goes primarily through an efficient risk analysis and vulnerability assessment that focuses on most critical resources in the system and highlights the security flaws to be considered later when the network is effectively designed.

1.2. On the impact of risk assessment in the decision-making process for network security designers

This very general picture of the ISSRM process highlights the importance of the intermediate risk assessment step, generally considered as the nucleus of risk management processes' lifecycle. As a matter of fact, next steps, such as security controls and countermeasures implementation, highly depend on the success of the risk assessment step. For instance, if the risk has been overestimated, administrators will likely implement high-cost protection devices to mitigate a risk that actually necessitates cheaper equipments.

Many approaches can be used to evaluate the risk related to information and network security systems. Most often, the security risk is expressed as:

$$Risk = Likelihood * Impact \qquad [1.1]$$

Indeed, risk assessment is usually conducted on the basis of threat likelihood and impact, which are, respectively, the probability of a threat occurrence and its potential damages to the system. A threat is defined as the possibility for an intruder to attack a system by exploiting existing vulnerabilities. However, this is one general expression among many, and as involved factors (e.g. likelihood, impact) could be modeled in many ways, numerous security risk assessment methods have already been proposed as described in section 1.3 and in Chapter 2.

1.3. Quantitative versus qualitative risk assessment approaches

As we mentioned in step (c) of Figure 1.1, risk assessment techniques can be undertaken in a quantitative or qualitative approach.

Typically, qualitative risk assessment relies on security specialists expertise and, most of the time questionnaires are used to gather their opinions, like in [BEN 92]. This can be costly as security expertise is expensive for companies. Furthermore, a data collection process is considered complex as it requires much time and effort, and might induce some computation errors (because they are performed manually). Besides, the qualitative results are mostly based on a ranking scale, and cannot be substantially evaluated because of their subjective nature.

For instance, it is possible to compare two security risk levels (e.g. between high and low) but impossible to estimate the distance between these measures (e.g. between two levels ranked as high). Moreover, the security expertise is generally based on the expert's intuitiveness and past experiences in the field, which does not always reflect the current and real situation. Then, qualitative risk assessment techniques likely suffer from a lack of sound theoretical bases, which do not give concrete knowledge about the information security risk.

Quantitative risk assessment allows a more accurate analysis of risk events and, to some degree, solves the issues related to qualitative techniques. In fact, a plethora of parameters involved in the risk assessment process can be used and are designed in many ways owing to mathematical and theoretical models. For instance, some designers might focus on modeling the impact of threats on business assets, whereas others decide to concentrate their efforts on attack progression modeling using Petri's network [JIN 09]. This allows a sharper analysis of risk events compared to qualitative techniques.

Besides, the quantitative results are mostly accurate and can be expressed either in business or technical languages. Thus, this makes it easier for enterprises to reach their financial objectives. Furthermore, it could be helpful for administrators willing to enhance the security of their networks. Quantitative risk assessment methods are usually supported by automated tools, which have the advantage of accelerating the assessment process and avoiding some computational errors.

Furthermore, quantitative risk assessment techniques can be used either for a preventive risk analysis or a reactive risk analysis depending on the context of the study. A preventive risk analysis often relies on the annual loss expectancy (**ALE**) index [MIC 04], which is the annual monetary loss that can be expected by a company according to the identified likely risk events. From a financial point of interest, ALE is an important metric that can be used directly in a cost-benefit risk analysis.

Quantitative risk assessment techniques also support reactive analyses, which are generally conducted to identify security countermeasures when an alert corresponding to an attack is triggered by a monitoring system. This could be done using, for instance, a NIDS system. For this purpose, several decision criteria are used and modeled in various ways. The most prominent models are detection and reaction cost models (e.g. the number of security countermeasures to deploy, the percentage of intrusions into the supervised network, the monetary or processing resources required to face an attack) [BAR 09], attack models (e.g. scenarios-based or tree-based graphs) [WIN 08], and threat impact models (e.g. impact distribution laws, impact progression over the network) [LAO 08].

Succinctly, qualitative and quantitative information security risk assessment approaches could be compared from three points of view: subjectivity, efficiency and cost. Table 1.1 depicts a summary of the advantages (denoted by +) and drawbacks (denoted by –) of each approach according to these three axes.

Criteria	Quantitative Approaches	Qualitative Approaches
Subjectivity	– At a design level + Solid theoretical models + Several factors are modeled	– Security experts intuitiveness and past experiences – Pedestrian risk evaluation
Efficiency	+ Numerical risk estimation (comparison is always possible) + Automated procedures/ tools (less errors) – Based on advanced aspects (not adapted for beginners)	– Ranking scale (difficult to compare) – Computation errors (human in the loop) – Preventive/reactive analyses are difficult to conduct
Cost	+ Relatively fast (only time needed by the tool) + No extra-expense	– Time-consuming procedures (e.g. questionnaires collection) – Financially expensive

Table 1.1. *Qualitative versus quantitative risk assessment approaches*

Looking at the comparison made above, it makes sense to confirm that a quantitative risk assessment approach is strongly preferred. A lot of work has been done in this area and we provide further a summary of the major research in this field (section 2.2).

Consequently, our methodology will be based on a quantitative assessment of each parameter involved in the global risk processing. However, the survey of these quantitative risk assessment methods emphasized another point that should be considered when the so-called risk assessment methodology has to be designed for an information system network, namely the network security risk propagation.

1.4. Network security risk propagation concept

1.4.1. *Impact of node correlation*

To understand the importance of the network security risk propagation concept in the design of a risk assessment methodology, let us see what could be the simplest definition of the word "network". According to the online Cambridge Dictionary, it is "*a large system consisting of many similar **parts** that are **connected** together to allow movement or **communication** between or along the parts or between the parts and a control centre*"[1]. Starting from this definition, we can deduce three important concepts that must be considered carefully when a risk assessment method has to be designed for a network information system:

– *Nodes*: these are the main components of a network information system, such as end systems (terminals, servers) and intermediate systems (hubs, switches, gateways). Every node has its own set of vulnerabilities that can be related to hardware, software, protocol stack, etc.

– *Physical interconnection between nodes*: as we have seen in the definition, nodes are interconnected by physical supports in a network. For example, nodes can be interconnected using cables (shielded twisted pair cable for instance) in a wired LAN or radio waves in a wireless LAN (**WLAN**) such as WiFi.

– *Communication (i.e. data flows) between nodes*: some nodes are able to provide services (FTP (File Transfer Protocol) transfer, HTTP (HyperText Transfer Protocol) browsing, database access, etc.). When two nodes want to communicate together, they must be interconnected physically and logically.

1 http://dictionary.cambridge.org/dictionary/british/network_1

Considering all these factors, it is not easy to deduce exactly the total risk of a large network, even if we can evaluate this risk node-by-node. In fact, apart from individual vulnerabilities, the global network security can be seriously compromised by the interconnected nodes. Indeed, many endogenous and exogenous factors have to be analyzed in order to determine as accurately as possible the risk level for the whole network.

On the one hand, the global network risk can be very low even if the risk related to a single node is very high (e.g. this node is isolated from the rest of the network and does not communicate with many other nodes). On the other hand, the security of the whole network can be heavily compromised by nodes that have strong interconnections and data flow exchanges with the rest of the network, even if those nodes have individually a low network risk.

Therefore, a network security risk should no longer be evaluated individually, but rather globally taking into account the service dependencies and node correlation. The security risk propagation within an information system network consolidates this idea that network intrusions are likely transitive processes.

1.4.2. *Network security risk transitivity*

When an attack occurs on a network node, it is highly likely that the intruder will try to attack the interconnected nodes when this is allowed by the network topology. The attacker would be able to do so if there were some system assets that could help him to break into a connected node. These assets could be applications, services (intruded on the associated port), user logins (e.g. root privilege access) or database access accounts. Strong dependencies between these system facilities imply some kind of transitivity in the network risk propagation process.

By way of an example, let i and j be two correlated nodes in the network and t an exploitable vulnerability on node j as shown in Figure 1.2.

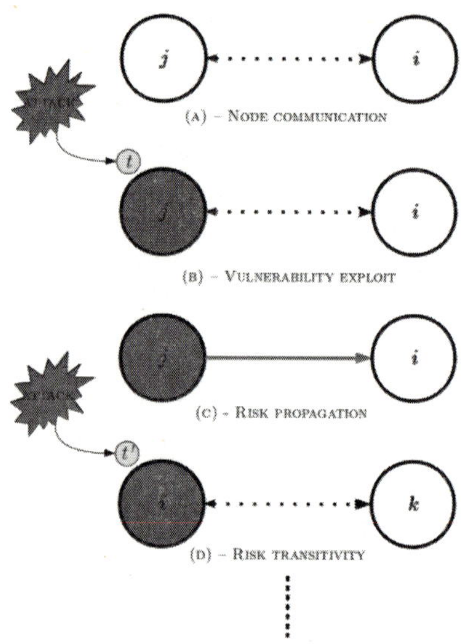

Figure 1.2. *Risk transitivity between correlated nodes*

Since node *j* has some vulnerabilities that could be exploited by an attack (step B), it might transmit its correlative risk to the connected node *i* (step C). This risk will propagate to the different nodes connected with node *j*. Besides, as long as the risk has been propagated from node *j* to the correlated node *i*, there is a strong probability that the intruder continues along his way and tries to break in to nodes connected with *i* (node *k* in step D). To provide a deeper understanding of the network security risk transitivity, we illustrate the risk propagation concept through a practical intrusion scenario in the following section.

1.4.3. *Network security risk propagation illustrative case*

Figure 1.3 illustrates an example of a step-by-step network security risk propagation into a simple LAN network. Let us say that administrator users on node A are allowed to log on a Web server (node B) using the Secure SHell (SSH) [YLO 06] service in order to manage a Website and refresh its content. Users possessing root privileges on node B are allowed to access a database (node C) that contains confidential data (e.g. Website user information like emails, credit cards, addresses). Only root users on node B are allowed to access the database on node C: for this purpose, a firewall (node D) is deployed and configured to filter the access to node C, meaning users from node A (even those with administrator privileges) are prevented from logging in to the database.

However, node A could suffer from a vulnerability that is still exploitable (i.e. not already fixed). An intruder may first exploit this specific vulnerability to node A (e.g. Operating System (OS) vulnerability) to get administrator privileges. He would probably face some issues trying to access directly node C from node A, but he could gain access to node B using the SSH service. In a second phase, he may try to grant root privileges on node B, then access the confidential data on node C without being intercepted.

In this section, we turned our attention to a second point of interest in network security risk assessment, namely the risk propagation. We showed that a risk should not be considered under a classical perspective (i.e. individually node-by-node), but instead at a higher level such that the impact of node correlation is taken into account in the risk computation.

The methodology presented later takes into consideration both quantitative assessment and risk propagation concepts. The proposed approach could help administrators willing to compare different security policies and find a cost-effective and secure policy. Besides, they will be able to evaluate the impact of any topological change in the network architecture (e.g. adding or deleting a node) on the network security. All the parameters involved in the network risk measurement will be explained and quantified: threat likelihood, risk impact (i.e. cost of damages), individual network risk (i.e. specific to a single node) and the total risk induced by the interconnection between the network components.

As the reader may notice, this methodology can be applied to any computer network and is not specific to a particular environment. Since we are able to quantify the logical interconnection between network nodes, the security assessment framework would fit to measure the risk on that network. While specific characteristics (such as priority between network domains) have been included to support data link communications, the aeronautical network remains the case study of the presented methodology as we will see in the dedicated section.

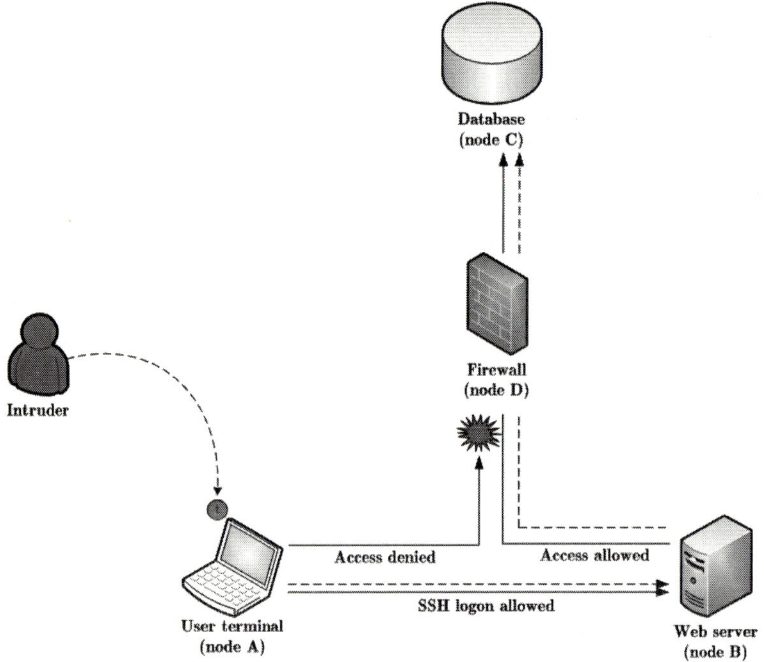

Figure 1.3. *Network security risk propagation example*

In Chapter 5 of this book, we apply our approach on the SESAR network architecture. In this context, we focused on the AeroMACS access network topology for airport communications. The goal was to discuss the risk results for the isolated AeroMACS scenario and to compare them regarding the intrinsic authentication/authorization security mechanisms in order to finally find which scenario holds the lowest network risk and to provide at the end some security guidances for future AeroMACS implementations.

The validation experiments relied on vulnerability statistics issued from the National Vulnerability Database (NVD)[2] and the Common Vulnerabilities and Exposures (CVE) database published by the National Institute of Standards and Technology (NIST). The NVD provides information about vulnerabilities such as type, class of severity and scores, extended descriptions, products or versions affected. Other vulnerability and statistical reports exist such as Secunia[3] or OSVDB (Open Source Vulnerability Database)[4] database: we picked the NVD database because it provides the Common Vulnerability Scoring System (CVSS) [SCH 05] severity score of a vulnerability, which is an essential quantitative parameter in our methodology.

2 http://nvd.nist.gov/

3 http://secunia.com/

4 http://osvdb.org/

2

Security Risk Management Background

ISSRM is a very active domain and many methods have already been proposed in the last decade. This chapter presents the different risk management methodologies that can be currently used (quantitative and qualitative **methodologies**). These methods are mainly driven by standards and guidance provided by professional and standardization organizations. In this chapter, the most-known security risk management methods in the security community are discussed. Indeed, there are more than 200 ISSRM methods in the literature; however, plenty of them are private and used in the scope of a single company or enterprise. In the literature, the main methods that are usually quoted are CCTA (Central Computer and Telecommunications Agency) CRAMM (CCTA Risk Analysis and Management Method), OCTAVE (Operationally Critical Threat, Asset, and Vulnerability Evaluation), EBIOS (*Expression des Besoins et Identification des Objectifs de Sécurité*), MEHARI (Harmonised Risk Analysis Method) and CORAS (Risk Assessment of Security Critical Systems). Even if several risk assessment methodologies have already been proposed in the literature, none of them can fully contribute to network attack complexity modeling. That is why in the next chapter we introduce a new quantitative risk assessment method able to take into account network risk propagation.

2.1. Qualitative security risk management methods

In this section, we discuss the most-known security risk management methods in the security community. Indeed, there are more than 200 ISSRM methods in the literature; however, plenty of them are private and used in the scope of a single company or enterprise. In the literature, the main methods that are usually quoted are: CCTA, CRAMM[1], OCTAVE [BEN 92], EBIOS [DCS 04], MEHARI[2] and CORAS [FRE 02].

2.1.1. *CRAMM*

The CRAMM method was developed by the CCTA of the United Kingdom in 1985. Currently, the method is owned by the SIEMENS company and is supported by a software tool that provides guidance to collect required data and adequately exploit the method. CRAMM is built around three steps: asset identification and evaluation, threat and vulnerability assessment, and countermeasures selection and recommendation. CRAMM is the only ISSRM method that explicitly recommends the use of quantitative risk assessment, but it uses pedestrian-scale values ranging from very low to very high.

2.1.2. *OCTAVE*

The OCTAVE risk-based security assessment method was published in 2001 by the Software Engineering Institute of the Carnegie Mellon University of Pittsburgh, USA. The method has the original approach of decomposing the risk assessment into three phases: building asset-based profiles, identifying infrastructures

1 http://www.cramm.com/

2 http://www.clusif.asso.fr/en/clusif/present/#mehari_link

vulnerabilities and developing security strategy and plans. OCTAVE aims to examine organizational and technological issues and define the best security strategy to face risk events.

2.1.3. *EBIOS*

EBIOS was created in 1995 by the French Central Directorate for Information Systems Security (DCSSI)[3] under the authority of the French Network and Information Security Agency (FNISA). EBIOS is assumed to support all the ISSRM processes (e.g. risk assessment, risk mitigation) and is strongly recommended for military and governmental information systems use. This method is still commonly used in the industry by several companies. EBIOS is composed of five steps: context of study, expression of security needs, threat study, identification of security objectives and identification of security requirements. EBIOS is also supported by an open-source software tool that makes the use of the method by non-initiated users easy.

2.1.4. *MEHARI*

MEHARI was created in 1996 by the *Club de la Sécurité de l'Information Français* (CLUSIF)[4,5], a French association of information security professionals. The MEHARI method is composed of several modules (security analysis and classification, evaluation guide for security services and risk analysis guide), which are centered around risk assessment and management processes. The advantage of this method is that each module can be used independently without

3 http://www.ssi.gouv.fr/en/

4 http://www.clusif.asso.fr/en/clusif/present/

5 CLUSIF contributes to information security education, improvements and awareness via publications resulting from the activity of its work groups, market studies and public conferences and meetings.

necessarily running the other modules. As for EBIOS, MEHARI is supported by a software tool[6].

2.1.5. *CORAS*

The CORAS ISSRM method results from a European project led by research institutes (e.g. the Queen Mary University of London) and commercial industries (e.g. the German Solinet[7]). The aim of the project was to develop a security framework supporting risk analysis and assessment for critical security systems (e.g. telemedicine and e-commerce applications). The method provides an integrated risk management and development process and an Extensible Markup Language (XML)[8] based platform.

2.1.6. *Discussion*

The approaches used in the risk management methods mentioned above evaluate damages produced by threats qualitatively, making results somewhat subjective (see section 1.3 for details).

Table 2.1 summarizes these methods and highlights the lack of a quantitative approach among the existing ISSRM methodologies.

In this perspective, the scientific community worked greatly on this specific point. In the next section, the main contributions made in the field of quantitative and formal representation of network security risk assessment are covered.

2.2. Quantitative security risk assessment approaches

There are lots of studies on network and information security risk quantification.

---

6 http://www.clusif.asso.fr/en/production/mehari/download.asp

7 http://www.solinet.com/

8 http://www.w3.org/TR/2008/REC-xml-20081126/

Standards and methods	Security oriented	Risk oriented	Aeronautical oriented	Quantitative approach
CRAMM	✓	✓	×	≈✓
OCTAVE	✓	✓	×	×
EBIOS	✓	✓	×	×
MEHARI	✓	✓	×	×
CORAS	✓	✓	×	×

Note: ✓, covered; ×, not covered; ≈✓, recommended but not used.

Table 2.1. *ISSRM standards and methods summary*

[CHE 06] outlines hierarchical threat assessment models for network security and quantifies the information system security parameters. Instead of focusing on quantifying the risk, this model focuses more on threats, meaning that the impact and probability of occurrence of potential vulnerabilities are assessed.

[DAN 09] prompts a service-based risk quantitative calculation method called service-based risk quantitative calculation (SRQC) for next-generation networks (NGN), which includes a layered risk assessment model (quantification of assets, vulnerabilities, threats and risk). The calculation model centers on NGN services and considers the relationship between services. The authors claim more valuable and objective results compared to the qualitative calculation. They proposed another three-dimensional security architecture model dedicated to NGN and focused this time on quantifying threat, vulnerability, stability and survivability parameters [HUA 09].

[YON 08] proposed a hybrid assessment method that combined qualitative analysis with quantitative calculation. The method is assumed to be ISO/IEC 27002 standard [ISO 05b] compliant and the authors used a multicriteria decision-making algorithm to weight the security parameters in the risk evaluation process. In the qualitative part of this research work, security experts have been asked to give different values to these weights, which bring us to the subjectivity- and qualitative-related issues.

[AHM 08] proposed a novel security metric framework combined with the NVD vulnerability database to identify and quantify objectively the most significant security risk factors: the existing vulnerabilities, the historical trend of known vulnerabilities, the prediction of potential new vulnerabilities and the associated impact severity. The authors made a considerable effort to model many specific risk parameters; however, the risk has been considered individually. The network risk propagation has not been included in the global network risk evaluation.

[LV 09] proposed a quantitative risk evaluation method for network security. The method analyzed the process usually used by attackers to intrude the network. The analysis was based on modeling attack activities and processes by tracking the transferring of safety states. Three key risk factors (assets, threats and vulnerabilities) have been identified and quantified. The attack probability indexes have especially been recurrently estimated by a layering approach based on the intruding process.

[MEN 09] proposed a risk evaluation method by formalizing and modeling attacks to find out how intruders proceed, and illustrated all the likely paths from their origins to their targets. The work is quite similar to [LV 09] as the authors also adopted a graph-tree-based attack modelization.

Some other quantitative studies used specific Bayesian network-based attack graphs to quantify the interdependency between existing vulnerabilities on each network node [FRI 08]. The authors proposed a probability model on attack graphs as a Bayesian network. The method provided the capabilities of using conditional probabilities to address the general cases of interdependency between vulnerabilities.

[KON 10] presented a causal model to encode logical relations inside the information system network. This contribution consisted in a probabilistic approach to model causal relationships among various threat sources and victim systems in order to facilitate quantitative and relational security assessment of information systems. The model is claimed to be context-free, meaning it makes no *a priori* assumptions regarding the application domain.

An approach for detection, assessment and prevention of further intrusions in a computer network has been presented in [YAU 99]. The approach used audit data from multiple network nodes and services. To achieve accurate results, inherent security relations among different network nodes have been considered: a security dependency relation (SDR) index has been defined to describe these relations, and a ripple effect analysis has been used to detect, assess and prevent intrusions based on SDRs. Several agents have also been used to improve the scalability and efficiency of the assessment approach.

[ZHA 04] proposed a risk assessment approach for network information systems based on node correlation, and an enhanced network node correlation (NNC) index has been introduced based on the SDR index presented in [YAU 99]. Finally, a quantitative taxonomy of a network node has been provided to support the quantitative calculation.

Table 2.2 summarizes the methods discussed above and how their authors managed to quantify the risk parameters.

The first thing noticeable from this quantitative information security risk assessment approach survey is that the authors provided different methods and formal techniques for generally reaching the same goal: assessing as accurately as possible the risk related to the network and information security. However, these approaches perform this task in different manners (e.g. service-oriented, threat-oriented, dedicated to specific networks like NGNs) and consequently have different coverage levels of the generic risk assessment process, which undeniably has an impact on the final risk scores. Each approach has its own positive points and weaknesses, depending on what perspective the user sees the network risk from.

For sure, if these methods are discussed from a network security risk propagation point of view, there is a blatant lack of consideration for this particular issue. Even when risk propagation has been introduced in a few works [FRI 08, KON 10, ZHA 04, YAU 99], some parameters such as the attack severity have been roughly quantified (or assessed using a qualitative technique). Most often, there is no indication on how the risk parameters are computed, which lead to some misunderstanding of the global assessment process. For instance, it has been pointed out in [ZHA 04] that the likelihood of the occurrence of threats is based on the analysis of attack frequency and complexity without any additional details. Also, the same authors claimed that attack impacts are assessed based on the analysis of severity results. They choose to let the administrators decide about the ranking of these parameters, which is totally subjective and at the opposite of quantitative risk assessment methods, baselines.

Approach reference	Quantitative assessment	Propagation oriented	Domain oriented	CVSS based
[CHE 06]	✓	×	×	×
[DAN 09]	✓	×	×	✓
[HUA 09]	✓	×	×	✓
[YON 08]	≈✓	×	×	×
[AHM 08]	✓	×	×	✓
[MEN 09]	✓	×	×	×
[LV 09]	✓	×	×	×
[FRI 08]	✓	✓	×	×
[KON 10]	✓	✓	×	×
[ZHA 04]	✓	✓	×	×
[YAU 99]	✓	✓	×	×

Note: ✓, covered; ×, not covered; ≈ ✓, hybrid approach (qualitative and quantitative analysis).

Table 2.2. *Quantitative risk assessment approaches summary*

2.3. Toward a quantitative propagation-based risk assessment methodology

Besides, network domain priority issues have simply been ignored. This is a surprising fact so far as several fields of applications in information and network systems deploy network domains and subdomains that do not necessarily have the same priority level or security needs (e.g. public network domain versus top secret or unclassified domain networks). A network domain is generally controlled by a single operator or an administrator who implements a uniform security policy within the domain. Hence, the level of security and the available security services are the same for all the nodes belonging to that network domain. When a network is composed of an interconnection of network domains, whether in a small network area (e.g. LAN) or a wider network (e.g. WAN), multidomain security should be considered. [SHE 06] is an example of a network multidomain security for IP multimedia subsystems. Another example is the Internet where cross-domain services are constantly provided. In such a case, depending on the delimited perimeter of the risk assessment study, the network node and the information carried should likely have a different value, meaning that the impact of a given attack will not be the same. Obviously, the aeronautical field is sensitive to these network domain priorities and should benefit from a methodology that has enough considerations regarding this particular requirement.

To address these issues (full quantitative assessment, network domain priority, risk propagation), we present a new quantitative methodology based on network security risk propagation in Chapter 3. To allow for a better integration of aeronautical network domain considerations throughout our approach, these concepts are taken into account at an early phase of the methodology design (see section 4.2 for details about aeronautical service instantiation for instance).

3

A Quantitative Network Risk Assessment Methodology based on Risk Propagation

This chapter introduces an original risk assessment approach based on risk propagation for network security. The proposed approach measures quantitatively the network risk level based on critical aspects such as the impact of a successful attack on a node and the risk propagation of that attack within the network. The experiments have been conducted using real data sets and vulnerability databases. Each parameter involved in the risk assessment process is quantified, then the overall approach is described in detail.

3.1. Quantifying methodology parameters

This section explains how every involved network parameter in the risk assessment process is computed. First, it is necessary to clarify how the risk is decomposed in the model, then every parameter is introduced with its respective formula.

3.1.1. *Network risk decomposition*

In the proposed methodology, there are four types of risks, namely:

1) *The risk per node* is computed for each node depending on its own vulnerabilities and its connections with correlated nodes. As a node is connected to other nodes in the network, the total risk for a given node *i* is evaluated as the product of node value $Value_i$ and the sum of its individual and propagated risks (respectively denoted $Risk_i^+$ and $Risk_i^-$):

$$Risk_i = Value_i * (Risk_i^+ + Risk_i^-) \quad [3.1]$$

2) *The individual risk* is the intrinsic risk computed for each node, which means that it takes into account only the vulnerabilities associated with the node itself. The individual risk $Risk_i^+$ is computed as the sum of the number of existing vulnerabilities T_i of the product between the likelihood of occurrence of a threat $P_t(i)$ and its impact $I_t(i)$, which is fully compliant with the basic expression of the risk mentioned in formula [1.1]:

$$Risk_i^- = \sum_{t=0}^{T_i} P_t(i) * I_t(i) \quad [3.2]$$

3) *The propagated risk* is the risk inherited from the dependency between correlated nodes (e.g. data flow exchanges, client–server architectures, etc.). The propagated risk $Risk_i^+$, is estimated as following:

$$Risk_i^+ = \sum_{j=0}^{n_i} \sum_{t=0}^{T_j} P_t(i,j) * I_t(i,j) \quad [3.3]$$

Compared to equation [3.2], the idea is quite the same except the difference that the propagated likelihood $P_t(i,j)$ and the propagated impact $I_t(i,j)$ are induced by all the vulnerable nodes connected with node *i*, where n_i denotes the total number of nodes connected to the node i.

4) *The network risk* is the total risk computed for all the nodes composing the network. It is calculated as the sum of all the risks

relevant to each node in the network (where n denotes the total number of hosts on the network):

$$Risk_{net} = \sum_{i=0}^{n} Risk_i \quad [3.4]$$

Section 3.1.2 explains how the node value used in formula [3.1] is computed.

3.1.2. *Node value*

Considering that network nodes do not have the same functionalities, it can be considered that their degree of importance in the network may vary. Thus, the value of a node *i* is given by:

$$Value_i = n_i * FunctionValue_i \quad [3.5]$$

Indeed, the node functionality $FunctionValue_i$, which reflects the importance of a host from a security point of view, is taken into consideration. For instance, it is clear that a firewall is more critical for the security of the network than a user terminal. Besides, the value of a node increases when it is connected to a large number of nodes (i.e. when the node correlation is high). For instance, server nodes (e.g. email servers, Webservers, DNS servers, etc.) or proxies are highly connected to other nodes in the network.

This node value expression could be sufficient if the risk assessment is dedicated to an intra-network domain perimeter where only the node functionality may vary from one host to another host. Nevertheless, the node value expression presented in equation [3.5] is certainly unsatisfactory to introduce the priority between network domains as discussed in section 2.3.

3.1.3. *Enhanced node value*

To introduce the priority between network domains, the equation [3.5] is slightly modified:

$$Value_i = n_i * FunctionValue_i * ClassValue_i \qquad [3.6]$$

where $ClassValue_i$ expresses the class value of a node, depending on the network domain it belongs to. The reader may notice that both node function and class values are the only parameters requiring a "*human in the loop*" because there are no means to quantify them in practice. It is usually the duty of the administrator to assign these values depending on his/her needs and objectives. For instance, if the monetary value is an important parameter in the risk evaluation process, he/she could instantiate the function values to the cost of each host. For such a purpose, the risk assessment tool must give enough freedom to the user that he can assign the node values as he/she wants before initiating the evaluation process. In the model, both functions and class values have been ranged between 0.0 and 1.0 (which is an arbitrary choice and could be modified). For the aeronautical case studies, $ClassValue_i$ and $FunctionValue_i$ are instantiated in section 4.2 for the specific needs of the Single European Sky ATM Research (SESAR) project.

In the following section, the individual and propagated impact of a threat on network nodes will be quantified.

3.1.4. *Impact of threats*

As seen in equations [3.2] and [3.3], two types of impact of threats have been defined:

1) $I_t(i)$ is the impact of threat caused by the exploit of a specific vulnerability *t* to the node *i*. For each node, the estimated impact $I_t(i)$ is retrieved from the CVE public vulnerability database. To be more specific, $I_t(i)$ is referred to the CVSS severity (i.e. impact) score

associated with each vulnerability occurring on that node, which is basically a numerical score ranged between 1.0 and 10.0. Practically, this information can be obtained by accessing manually the CVE database or by scanning the network using the NESSUS[1] vulnerability scanning tool: as far as NESSUS is connected to the CVE database, the results remain the same. The number of vulnerabilities T_i, used in both equations [3.2] and [3.3], is a simple addition on the existing vulnerabilities for that node.

2) $I_t(i,j)$ is the propagated impact of a threat caused by the exploit of a vulnerability *t* from a node *j* to a node *i*:

$$I_t(i,j) = Value_i * I_t(j) * \sigma(i,s) \quad [3.7]$$

The propagated impact depends on the affected node value: $Value_i$ (as expressed in formula [3.6]), the impact of *t* on the issuing node *j* (*see* CVSS score as defined above) and the targeted service s. $\sigma(i,s)$ is a scalar value deduced as the following:

$$\sigma(i,s) = SPV_i * (SOV_s)^{tr} \quad [3.8]$$

where SPV_i is a binary *Security Protection Vector* that defines the security features provided by security mechanisms and countermeasures used to protect node *i*. The dimension of the vector depends on how many security services are provided. For instance, if confidentiality, integrity and authentication are considered, SPV_i would be a three-dimensional vector. Moreover, let us assume that only confidentiality is provided for the data flows issued from the node *i*, the associated binary indicator vector will be equal to [0 1 1]. It could seem meaningless to associate the zero binary value to express a "*YES*", but this is done in order to respect the impact function behavior. Indeed, the impact grows when less security features are available. Thus, mapping the one binary value to a "*YES*" is inadequate. In this specific case, the more security features there are, the bigger would be the propagated impact, which would not be logical regarding the impact function previously defined.

1 Note that any scanning tool providing the same features as NESSUS can be used.

The second part of equation [3.8] is the *Security Objective Vector* SOV_s that defines the security objectives per service (note that the transpose of the vector is used here in order to obtain a scalar result). For instance, if the example of a service *s* provided by the node *i* is chosen again where high security objectives are expressed for the confidentiality, integrity and authentication security services, the SOV_s vector could be equal to [5 5 5], where the value 5 expresses the highest security requirement. These values depend on the security objectives expressed beforehand at a previous step of the ISSRM process (see step (b) of Figure 1.1).

In section 4.2 the SOV_s vector is instantiated according to the security objectives expressed in the Communications Operating Concept and Requirements (COCR) document for the aeronautical operational services.

In the following section, explanations are given on how to quantify the occurrence likelihood and propagation of threats.

3.1.5. *Likelihood of threats*

Similar to the impact of threats, two types of threat likelihood are defined:

1) $P_t(i)$ is the occurrence likelihood of a threat caused by the exploit of a specific vulnerability *t* to the node *i*. $P_t(i)$ represents the possibility that an attack associated with a specific vulnerability *t* is conducted. The occurrence likelihood evaluation is driven by the Telecommunications and Internet Protocol Harmonization Over Networks (TIPHON) threat analysis methodology [ETS 03] proposed by the ETSI. However, as the likelihood values are qualitative, this part of the existing European Telecommunications Standards Instituate (ETSI) methodology is extended in order to quantify the involved parameters. Indeed, as described in [ETS 03], the likelihood evaluation is based on two behavioral factors:

- The technical difficulties that an attacker may face in order to achieve his/her goal.

- The motivation for an attacker to carry out a given attack.
As it appears in [ETS 03], the likelihood of threats can be:

a) *Unlikely*: if the motivation for conducting an attack is low (e.g. no financial interest or technical challenges) and there are strong technical difficulties to overcome (e.g. insufficient knowledge to conduct the attack);

b) *Possible*: if the motivation is moderate (e.g. reasonable financial gains) and the technical difficulties are solvable (e.g. information required to exploit the vulnerability are available);

c) *Likely*: if there is a high attacker motivation (e.g. inducing a denial of service on the network, important financial gains) and technical difficulties are almost non-existent (e.g. no security protection).

Despite the fact that these behavioral factors seem adapted and should be logically considered to evaluate the likelihood associated with an attack, they cannot be used directly in the model because quantitative values are needed. Besides, the ETSI methodology does not explain how the technical difficulties and motivation factors are combined together in order to deduce the likelihood of the occurrence of a threat. To address these issues, some modifications are brought in order to use these behavioral factors in the risk assessment process. First, the likelihood is computed using the motivation and technical difficulties values (respectively denoted $Motivation_t(i)$ and $TechnicalDifficulty_t(i)$) as shown in the following formula:

$$P_t(i) = \frac{Motivation_t(i)}{TechnicalDifficulty_t(i)} \quad [3.9]$$

In fact, $P_t(i)$ should increase when the motivation gets higher; otherwise, $P_t(i)$ decreases when the technical difficulties that must be overcome increase.

The motivation for an attacker to exploit a vulnerability *t* on a node *i* is:

$$Motivation_t(i) = Value_i * T_i \quad [3.10]$$

Equation [3.10] shows that the motivation increases as the node value or the number of known vulnerabilities increases. Technical difficulties become more significant when security features (e.g. firewalls) are reinforced (e.g. increasing their number or enhancing the security policies) or the amount of information required to exploit a vulnerability *t* is high:

$$TechnicalDifficulty_t(i) = S_i + B_t \qquad [3.11]$$

In equation [3.11], S_i expresses the number of security mechanisms used to protect the node *i*. B_t is the amount of elementary informations needed by an attacker to exploit the vulnerability *t*. The assumption that $B_t > 0$ is reached, means that at least one elementary information has to be available in order to conduct an attack exploiting *t*. Indeed, the attacker will probably be unable to exploit a vulnerability if a minimum of data is not available to start the attacking process (e.g. opened port IDs, user logins, target addresses, etc.). As the resulting probability value must be ranged between 0.0 and 1.0, both motivation and technical difficulties values have been normalized between 0.0 and 1.0.

2) $P_t(i, j)$ is the likelihood of the propagation of a threat caused by the exploit of a vulnerability *t* on a node *j* to node *i* given by:

$$P_t(i, j) = P_t(j) * P(i, j) \qquad [3.12]$$

In fact, the likelihood of propagation depends on the likelihood of vulnerability *t* on the issuing node *j* and the likelihood of correlation $P(i, j)$ between the two nodes, given by:

$$P(i, j) = \frac{f_{ij}}{F_{ij}} \qquad [3.13]$$

$P(i, j)$ depends on the ratio between the number of detected and total data flows exchanged (i.e. the sum of all detected data flows) between two nodes *i* and *j* and relative to the attacked service. Practically, f_{ij} and F_{ij} can be directly deduced using some network statistics tools like NETSTAT[2] or raw data from */proc/net/dev*.

2 http://linux-ip.net/html/tools-netstat.html

All the parameters discussed above are summarized in Table 3.1:

Notation	Description
$Risk_i$	Node risk evaluated on node *i*
$Risk_i^-$	Individual risk evaluated on node *i*
$Risk_i^+$	Propagated risk evaluated on node *i*
$Risk_{net}$	Network risk
$Value_i$	Value of node *i*
$FunctionValue_i$	Function value of node *i*
$ClassValue_i$	Class value of node *i*
t	An exploitable vulnerability
n	Total number of nodes in the network
n_i	Number of nodes connected with node *i*
T_i	Number of vulnerabilities exploitable on node *i*
S_i	Number of security features deployed to protect *i*
B_t	Number of information needed to exploit *t*
$P_t(i)$	Likelihood of an occurrence of a threat exploiting *t*
$I_t(i)$	Impact of threat exploiting *t*
$P_t(i,j)$	Likelihood of the propagation of a threat exploiting *t*
$I_t(i,j)$	Propagated impact of a threat exploiting *t*
f_{ij}	Number of flows detected between nodes *i* and *j*
F_{ij}	Total number of flows exchanged between *i* and *j*
$Motivation_t(i)$	Motivation of an attacker to exploit *t*
$TechnicalDifficulty_t(i)$	Technical difficulty level to exploit *t*
$\sigma(i,s)$	Scalar value as the product of SPV_i and SOV_s
SPV_i	Security protection vector for the node *i*
SOV_s	Security objective vector for the service *s*

Table 3.1. *Risk parameter notations*

Finally, because all the parameters involved in the network risk computation process are now defined, the risk assessment algorithm can be proposed. Each step of the overall network risk assessment process is explained regarding the relevant pseudocode algorithm.

3.2. Network security risk assessment process

In this section, six steps are described leading to the final network risk evaluation based on the proposed approach. In each step, the corresponding pseudocode algorithm is given:

1) *Initiation step*: First, a set of vulnerable nodes and a set of processed nodes to null are initiated. All risk values (individual, propagated and node risks) are initiated for each node in the network. Also, a set of correlated nodes for each node is created.

Algorithm 1. *Variables initiation*

$V \leftarrow \{\varnothing\}$; *//initiate a set of vulnerable nodes*
$NV \leftarrow \{\varnothing\}$; *//initiate a set of processed nodes*
for all $i \in network$ **do**
    $Risk_i^- \leftarrow 0$;
    $Risk_i^+ \leftarrow 0$;
    $Risk_i \leftarrow 0$;
    $C_i \leftarrow \{\varnothing\}$; *//initiate a set of correlated nodes with node i*
end for

2) *Scan and identify vulnerable nodes*: The second step is to identify all the vulnerabilities specific to each node in the network. If any vulnerability is detected, the node is marked as vulnerable and added to *V*. The node values are also computed and the corresponding CVSS severity scores are stored for a later use in the algorithm (see algorithms 3 and 4).

3) *Compute the individual risk for each vulnerable node*: For each vulnerable node identified in the previous step, tracks of the nodes that could be impacted by the network correlation are kept. The obtained set of correlated nodes will be used in the following step. Then for each vulnerable node, its individual risk is computed according to the formula [3.2].

4) *Compute the propagation risk for nodes correlated with vulnerable nodes*: For each vulnerable node, nodes are considered one-by-one, and the propagated risk of the targeted node is incremented.

The infected node is then tested: if it was not considered as vulnerable, the set of vulnerable nodes and the occurrence likelihood of threat are updated.

Algorithm 2. *Scan and identify vulnerable nodes*

for all $i \in network$ **do**
    identify vulnerabilities;
    $Value_i \leftarrow n_i * FunctionValue_i * ClassValue_i$;
    if any vulnerability is detected **then**
        add node i to V;
    end if
    for all vulnerability t **do**
        store t and associated *CVSS* score;
    end for
end for

Algorithm 3. *Compute the individual risk for each vulnerable node*

for all node $i \in V$ **do**
    store correlated nodes with node i in C_i;
    for all vulnerability t **do**
        $TechnicalDifficulty_t(i) \leftarrow S_i + B_t$;
        $Motivation_t(i) \leftarrow Value_i * T_i$;
        $P_t(i) \leftarrow Motivation_t(i) / TechnicalDifficulty_t(i)$;
        $Risk_i^- \leftarrow Risk_i^- + (P_t(i) * I_t(i))$;
    end for
end for

5) *Compute the risk for each node in the network*: At this phase, the output of algorithms 3 and 4 is considered and the risk per node is computed according to expression [3.1].

6) *Compute the whole network risk level*: Finally, the total network risk is estimated according to formula [3.4].

Algorithm 4. *Compute the propagation risk for nodes correlated with vulnerable nodes*

while $V \neq \{\varnothing\}$ **do**
  for all $j \in V$ **do**
    for all $i \in C_j$ **do**
      for all vulnerability *t* **do**
        $\sigma(i, s) \leftarrow SPV_i * (SOV_s)^{tr}$;
        $I_t(i, j) \leftarrow Value_i * \sigma(i, s) * I_t(j)$; *//s is the targeted service by t*
        $P(i, j) \leftarrow f_{ij}/F_{ij}$;
        $TechnicalDifficulty_t(j) \leftarrow S_j + B_t$;
        $Motivation_t(j) \leftarrow Value_j * T_j$;
        $P_t(i) \leftarrow Motivation_t(i)/TechnicalDifficulty_t(i)$;
        $P_t(i, j) \leftarrow P_t(j) * P(i, j)$;
        $Risk_i^+ \leftarrow Risk_i^+ + (P_t(i, j) * I_t(i, j))$;
        $P_t(i) \leftarrow P_t(i) + P_t(i, j)$; *//update the likelihood of threat*
        if $P_t(i) > 1$ **then**
          $P_t(i) \leftarrow 1$; *//the likelihood of threat should not exceed 1*
        end if
      end for
      if node $i \notin V and \notin NV$ **then**
        store node *i* in *V*; *//the node is now vulnerable*
      end if
    end for
    copy node *j* to *NV* and remove it from *V*; *//this node has been processed*
  end for
end while

Algorithm 5. *Compute the total risk for each node in the network*

for all node $i \in network$ **do**
  $Risk_i \leftarrow Value_i * (Risk_i^- + Risk_i^+)$;
end for

Algorithm 6. *Compute the whole network risk level*

for all node $i \in network$ **do**
    $Risk_{net} \leftarrow Risk_{net} + Risk_i$;
end for

3.3. Conclusion

The first part of this book was dedicated to risk assessment background in network security and to the quantitative propagation-based methodology provided in Chapter 2. To validate the presented approach, a specific aeronautical network case study has been considered in the scope of the SESAR project. The security of the aeronautical mobile airport communication system (AeroMACS) technology used for airport communications between aircraft/surface vehicles and operational entities (e.g. air traffic controllers) has been discussed and then improved through the original risk methodology introduced in Chapter 3. As the reader may not necessarily have an aeronautical technical background, the second part of the book introduces the basic concepts of future data-based aeronautical communications with a focus on the AeroMACS network architecture and protocol stack.

PART 2

Application to Airport Communication Network Design

4

The AeroMACS Communication System in the SESAR Project

A specific aeronautical background is mandatory to fully understand the case study introduced in Chapter 5 of this book. Thus, specific aeronautical concepts and definitions are introduced in this chapter. First, the scientific context is introduced through a brief overview of the European project SESAR. Then, an aeronautical document-related communication operating concept and requirements entitled COCR used in the case study is considered. Speaking of which, the AeroMACS defined in SESAR 15.2.7 working package is considered. This chapter ends with an overview of AeroMACS technology and specifically focuses on access network vulnerabilities.

4.1. Overview of the European SESAR project

The European collaborative project entitled SESAR aims to modernize the future European ATM [EUR 11]. The project is still in progress and probably considered as one of the most important European R&D (Research and Development) collaborations ever launched by the European Commission, EUROCONTROL, TEN-T EA

(Trans-European transport Network Executive Agency)[1], and other actors from the industry such as Airbus[2] and Thales[3]. The aim of the project is to offer technical and operational solutions to meet future air traffic capacity and air safety needs. The total estimated cost of the development phase of SESAR is 2.1 billion euros, to be shared equally between the academic community, EUROCONTROL and industry.

Note that a similar project called NEXTGEN (Next Generation Air Transportation System) is being undertaken by the Federal Aviation Administration (FAA)[4] to improve the American National Airspace System (NAS) and deal with the same air traffic congestion issues [FAA 11].

4.2. Overview of aeronautical communications operating concept and requirements

To cope with the forecasted increased density in the worldwide airspace, EUROCONTROL and the FAA combined their efforts to identify candidate data-based technologies able to satisfy safety and regulatory requirements for Air Traffic Services (ATS) and Aeronautical Operational Communications (AOC). In 2002, a document entitled Communications Operating Concept and Requirements (COCR) [EUR 02] was produced as a result of this collaboration. Basically, the document covers communication systems and operational requirements, referred to by Future Radio System (FRS).

ATS and AOC operational voice services are briefly mentioned but the document clearly focuses on emerging data services, with a particular emphasis on message characteristics and specifications (e.g. message size, frequency of use per sector). The COCR document also covers safety and security requirements for operational data services

1 http://tentea.ec.europa.eu/

2 http://www.airbus.com/

3 http://www.thalesgroup.com/

4 http://www.faa.gov/

using an Operational Safety Assessment (OSA) methodology. These data have been useful for the SESAR project case study developed at the end of this book.

The COCR considers both current and future ATM specificities. Thereby, the study is divided into two phases:

1) *Phase 1* started already around 2005 in some regions of the world. In this phase, analog voice continues to be used for operational services whereas digital data serves as a backup for air–ground communications until 2030.

2) *Phase 2* is planned to start around 2020. In this phase, operational data services are expected to be the primary means of ATM communications, voice will be used only for exceptional circumstances or for areas that do not have data communications implemented yet.

Figure 4.1 illustrates the COCR phase 1 and 2 expected time frame [EUR 02]:

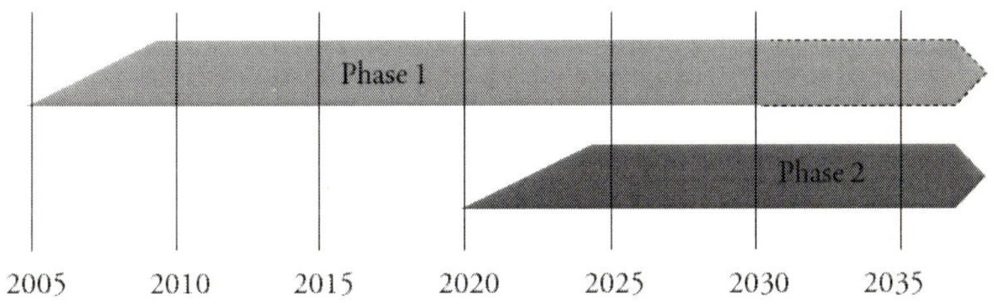

Figure 4.1. *COCR phases 1 and 2 concept evolution over time*

Besides, the document considers the following five representative airspace domains in the specification of operational data services (e.g. frequency of use of a given service per domain, total number of Air Traffic Control (ATC) ground stations deployed for each domain, support of voice and/or data messages, etc.):

1) *APT*: The APT (Airport) domain is the closest airspace to the airport surface. Its consists of a 10 miles cylindrical area and up to nearly 5,000 ft in height.

2) *TMA*: The Terminal Maneuvering Area (TMA) domain is the surrounding airport airspace with a typical 50 nautical miles (NM)[5] radius from the airport center. The TMA domain starts at 5,000 ft and ends at 24,500 ft. The COCR assumes the same characteristics for departure and arrival TMA domains.

3) *ENR*: This is the airspace surrounding the TMA domain and starting at FL245 to FL600. En Route (ENR) is the continental/domestic airspace when the airplane is cruising.

4) *ORP*: This airspace domain is identical to the ENR domain, the only difference is that oceanic, remote and polar (ORP) is associated with oceanic and unmanned areas. The ORP domain has a horizontal limit extending 1,000 NM by 2,000 NM.

5) *AOA*: This airspace domain will be operational starting from 2020. The Autonomous Operations Area (AOA) domain concerns autonomous operations where aircraft is able to make self-separation without any ATC directives. The COCR assumes that the AOA airspace domain has horizontal limits of 400 NM by 800 NM.

Figure 4.2 illustrates the organization of these five domains in the airspace as defined in the COCR document:

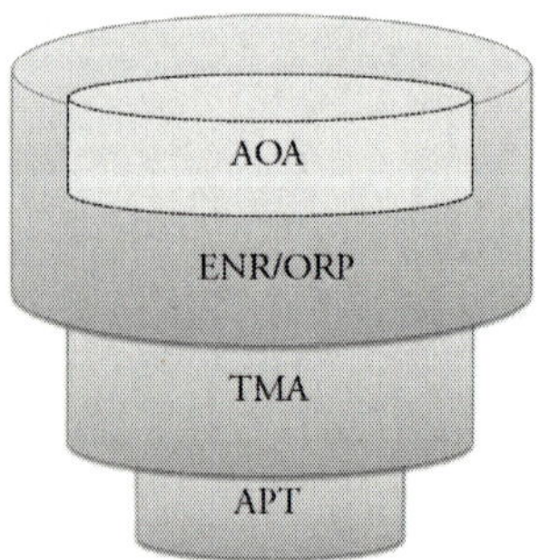

Figure 4.2. *General airspace decomposition*

5 In aeronautics, the Flight Level (FL) is expressed in feet. For example, FL800 corresponds to 80,000 ft. One foot (ft) is equal to 0.3045 meters (m). The nautical mile is used in both aeronautics and maritime navigation. One nautical mile corresponds to 1,852 m.

4.3. Introduction to the AeroMACS communication system

AeroMACS is a new aviation-dedicated transmission technology based on the WiMAX IEEE 802.16e standard[6]. The aim is to support fixed and mobile ground–ground and air–ground data communications at the airport surface today based on wifi architecture. The AeroMACS technology will allow mobile stations (MSs) such as aircraft and surface vehicles to communicate with airline operators and airport staff at three different surface zones: RAMP (where the aircraft is at the gate before departure), GROUND (the aircraft is taxiing to the runway) and TOWER (until the aircraft takes off). These three zones constitute the APT domain previously introduced.

Using a WiMAX-based technology standard is profitable for the aviation industry for many reasons. First, the standardization and deployment processes are fast and cost-effective at the opposite of a newly developed standard for the sake of airport communications. Moreover, the scientific community has been working on IEEE 802.16 standards for many years: highly qualified certification agencies such as the WiMAX Forum[7] are continuously looking after interoperability and technical issues related to the standard. The AeroMACS standard is currently a hot topic in datalink communications and many tests are already running this for a future deployment. For instance, an AeroMACS profile was recently developed jointly by the RTCA SC-223[8] and EUROCAE WG-82[9] and is intended to provide performance requirements for the system implementation.

By way of an example, Figure 4.3 shows a testbed AeroMACS network layout on the Cleveland-Hopkins International Airport in Ohio, USA, as it appears in [HAL 10]. Several subscriber stations

6 http://standards.ieee.org/about/get/802/802.16.html

7 http://www.wimaxforum.org/

8 http://www.rtca.org/comm/Committee.cfm?id=133

9 http://www.eurocae.net/working-groups/wg-list/50-wg-82.html

(SSs)[10] and base stations (BSs) have been deployed and tested by the NASA Glenn Research Center[11] for the sake of the NextGen project under the supervision of the FAA. BSs are responsible for providing air interface and additional functions (e.g. DHCP proxies, handoff triggering, etc.) to these MSs.

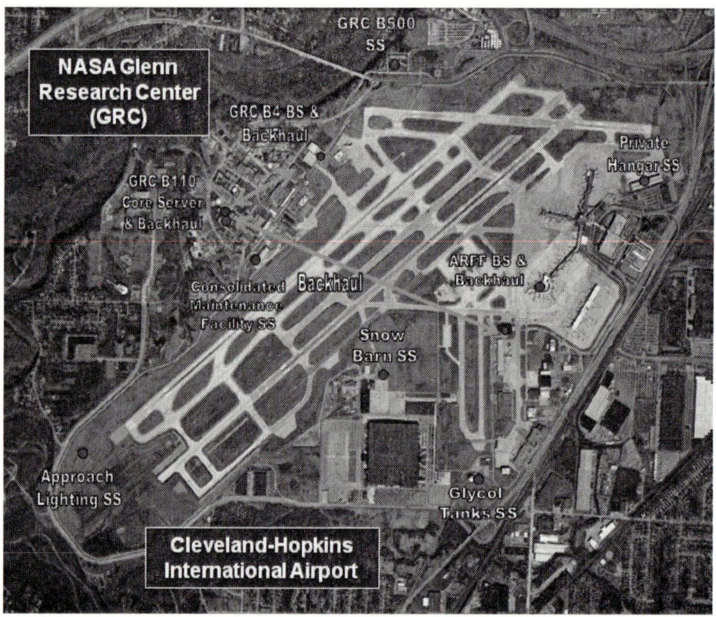

Figure 4.3. *Cleveland-Hopkins International airport AeroMACS testbed [HAL 10]*

4.3.1. *AeroMACS protocol stack*

The AeroMACS protocol stack is composed of two main layers: the Physical (PHY) layer and the Medium Access Control (MAC) layer, which are composed of three sub-layers as shown in Figure 4.4:

---

10 SS is the initial term used in the early versions of the WiMAX standard: after the IEEE 802.16e amendment, the MS term is being used to refer to station mobility.

11 http://www.nasa.gov/centers/glenn/home/index.html

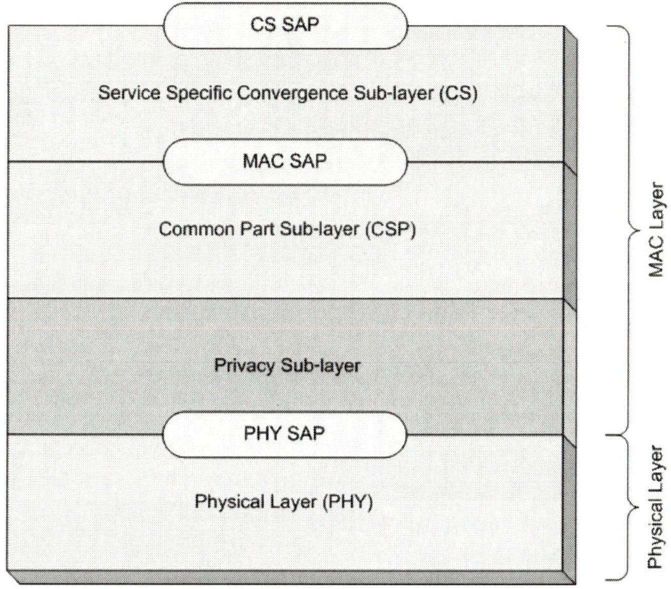

Figure 4.4. *AeroMACS protocol stack*

The first layer is the service-specific convergence sub-layer (CS) that communicates with higher layers through the CS Service Access Point (SAP), it acquires external network data and transforms them into MAC segment data units (SDUs). The second layer is the common part sub-layer (CPS) responsible for the system access, bandwidth allocation, connection management and MAC SDUs fragmentation into MAC protocol data units (PDUs).

The security is handled at the privacy sub-layer. It addresses many security services such as authentication, authorization, key establishment and encryption/decryption of data between the PHY and MAC layers. In fact, after the failures that restricted the early IEEE 802.11 networks, security has been seriously considered in the WiMAX standards (and consequently AeroMACS) and was built in (rather than built on) to the 802.16 protocol specifications since the first standard release in 2001. The privacy sub-layer provides several mechanisms designed to protect both service providers and users/customers from unauthorized access or information disclosure. However, the details of these security features are out of the scope of this book, but roughly speaking, the standard uses a privacy key management (PKM) protocol for secure key distribution and X.509 certificates for BSs and MSs identification. Security connections are maintained using security associations (SAs), advanced algorithms such as AES (for encryption), extensible authentication protocol (EAP) [ABO 04] or Rivest, Shamir and Adleman (RSA) (authentication, authorization) are used to protect the data exchanges.

The specification of the AeroMACS standard focuses on the physical and the access layer of the radio link as shown in Figure 4.4. However, this is not sufficient for a built-in broadband wireless airport network: end-to-end services such as QoS, security, mobility management and IP connectivity are a requisite and should be provided beyond the AeroMACS scope (i.e. at higher layers of the protocol stack). In this context, an AeroMACS network reference model has been developed.

4.3.2. *AeroMACS reference network architecture*

To understand the interconnection concepts and topology considerations made in the scope of the SESAR project, a logical representation of an AeroMACS reference network must be introduced first. Such a scheme distinguishes between the logical domains, the functional entities and the reference points (RPs) as reported in Figure 4.5. The main functional entities depicted are:

– MSs and SSs that could be, for instance, aircraft or surface vehicles.

– The Access Service Network (ASN) represents the boundary for functional interoperability between MSs and AeroMACS connectivity services. The ASN integrates many functions such as forwarding AAA messages between MSs and the Home Network service Provider (H-NSP), relaying network services messages (e.g. DHCP request/response), etc.

– The Connectivity Service Network (CSN) network provides connectivity to public networks such as the Internet.

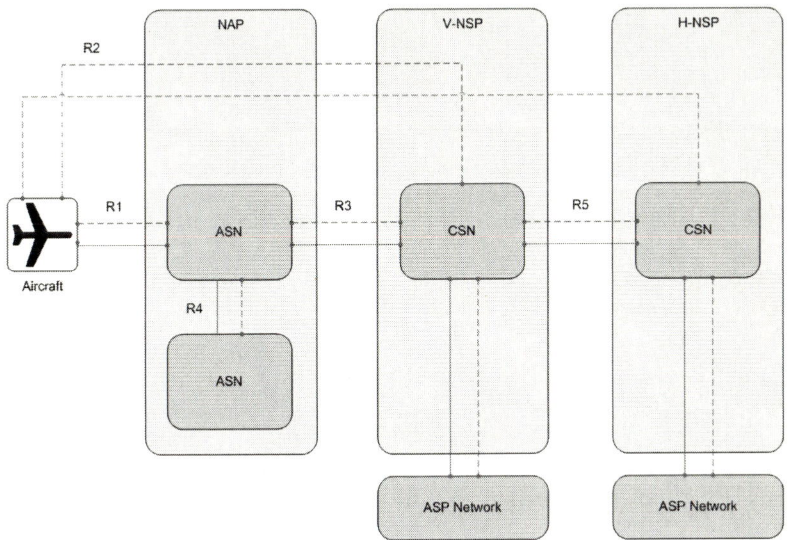

Figure 4.5. *AeroMACS network reference architecture*

The logical domains, which are basically the set of functions associated to a single domain, and considered in the network reference architecture are:

– The Network Access Point (NAP) is the physical point used by MSs to access the network.

– The H-NSP is the AeroMACS service provider that provides Service Level Agreement (SLA) to the MSs such as IP connectivity and core network services. These NSPs could be for instance, SITA, ARINC or even the airlines depending on the provided service.

– The Visited Network service Provider (V-NSP) is visited by the MSs to access the network in a roaming scenario (which usually depends on the roaming agreement made between the MS's H-NSP and the V-NSP).

Referring to Figure 4.5, RPs are the communication end points between functional entities and represent the interfaces that ensure the interoperability between AeroMACS-related components. Table 4.1 gives a description of all the reference network architecture interfaces:

RP	Interface	Functionality
R1	Between MSs and BSs	Air interface
R2	Between MSs and the CSN	IP host configuration, IP mobility, authentication, authorization
R3	Between the ASN and the CSN	Mobility management, authentication, authorization, tunneling
R4	Between ASNs	Mobility management, ASNs interworking
R5	Between CSNs	Roaming and interworking between the V-NSP and the H-NSP

Table 4.1. *AeroMACS reference point interface description*

The AeroMACS network reference model provides a unified and flexible network for airport data communications. However, because of the sensitive nature of operational data and the priority issues in aeronautical communications (generally, ATS have higher priority compared to AOC), the connected AeroMACS network should provide enough security as, for instance, data sent over the network remain secure.

4.3.3. *AeroMACS security considerations*

The privacy sub-layer has been defined to deal with several attacks targeting the authentication the authorization, or the confidentiality of exchanged data. Despite these security mechanisms, many security weaknesses have been discovered. Attacks ranging from physical layer-based attacks (e.g. jamming, scrambling, etc.) [BAR 05], attacks on the key management (e.g. mutual authentication, PKM failure, replay attacks) [XU 06], brute force attacks on privacy (e.g. Digital Encryption Standard (DES) in Cipher Block Chaining (CBC) mode is now considered insecure after operating $2^{\frac{n}{2}}$ blocks where n is the block size) [JOH 04], attacks on availability (e.g. DoS on PKM authorization phase) [ERN 08] have been identified. To mitigate them, many amendments have been conducted and the privacy sub-layer has been redefined for the mobile IEEE 802.16e standard. Still, some of these issues exist and are exploitable on the network as is explained in the next subsection.

Moreover, an end-to-end connected network (e.g. adding application servers, service provider nodes, etc.) induces an additional risk due to the node correlation and service dependencies as has been previously discussed. Thus, the next section focuses on the inherited AeroMACS vulnerabilities from the mobile version of the WiMAX standard. These security issues are mainly related to the integrity of some management communication messages and the key sharing scheme in multicast and broadcast services.

4.3.3.1. *Analysis of AeroMACS security weaknesses*

The AeroMACS security weaknesses can be grouped into three categories as follows:

– *Unauthenticated management messages:* most of the management messages defined in AeroMACS are integrity protected using a Hash-based Message Authentication Code (HMAC) [KRA 97] or alternatively a Cipher-based Message Authentication Code (CMAC) [NIS 05]. However, some messages (e.g. MOB_TRF-IND, MOB_NBR ADV, RNG REQ) are not authenticated which leads to some vulnerabilities. Also, some management messages are sent over the broadcast management channel: authenticating these types of messages becomes difficult because there is no common key to generate the message digest. Furthermore, a common key would not completely provide message integrity as mobile stations sharing the same key can arbitrarily forge these management messages and generate false authentication digests.

– *Unencrypted management messages:* when initial network entry procedure begins between a MS and a BS, many management messages (e.g. mobility parameters, power settings, security capabilities) are sent clearly. An adversary may eavesdrop on these messages just by listening to the channel in order to establish detailed profiles of MS or BS. More specifically, when Multicast and Broadcast Rekeying Algorithm (MBRA) is used, group encryption keys called Group Trac Encryption Keys (GTEKs) are distributed to all group members and encrypted using another key, namely Group Key Encryption Key (GKEK). This GKEK symmetric key is shared and known by all groups which means that a malicious group member may use a new GKEK key to update a request for the GTEK key and generate its own GTEK.

– *Shared keys in multicast and broadcast services:* multicast and broadcast messages are encrypted and authenticated using a symmetric shared key between a BS and all MSs belonging to the same group: this is an issue in the sense that any MS may impersonate the original BS by forging false multicast or broadcast messages and sending them to other MSs in the same group.

These security weaknesses have already been under inspection in recent works [HAL 10, NAS 09, WIL 11]. They have been clearly identified and the relevant security risks should be quickly fixed. Moreover, the network risk assessment methodology presented in this book is a valuable requisite in the sense that it could help us to determine which network topology or security policy (i.e. with the lowest network risk) provides the most secure system.

Section 4.3.4 depicts the AeroMACS network topology defined in the scope of the SESAR project. The different experimental results are discussed then, a subscenario is presented in order to compare two authentication/authorization security protocols supported by the AeroMACS privacy sub-layer, namely RSA (an algorithm which is a public key encryption technology) and EAP, in terms of vulnerability, individual and propagated risk.

4.3.4. *AeroMACS reference network topology*

4.3.4.1. *Isolated AeroMACS network architecture*

Figure 4.6 depicts the AeroMACS network topology, three main portions can be identified, namely the ASN, the CSN and the mobile stations (aircraft or surface vehicles).

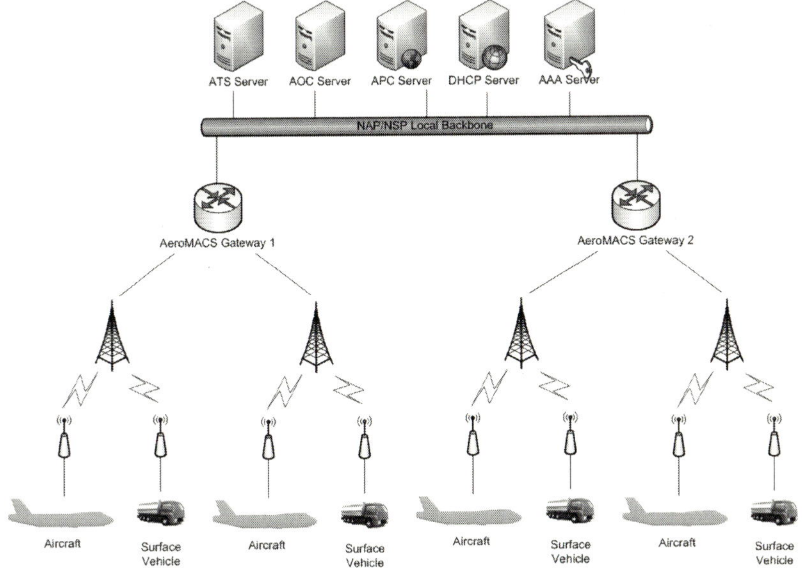

Figure 4.6. *Isolated AeroMACS network topology scenario*

This basic topology assumes the existence of a standalone service network supported by an AeroMACS access network in the airport. Consequently, all the services are provided by components inside the airport network (AAA, DHCP and application servers) and placed within the airport backbone.

The AeroMACS (additionally to the AAA server) segment is the only system supporting security features and the AAA server will be directly reachable through a dedicated gateway between the AeroMACS network and the Airport's other networks. The APC server in Figure 4.6 refers to the AirPort Communications server.

4.3.4.2. *End-to-end AeroMACS network architecture*

The end-to-end topology assumes that the AeroMACS (and the AAA server) segment are integrated in a global network infrastructure where other security features are deployed at several points (e.g. firewalls). In this configuration, there are two AAA servers: the home AAA server managed by the service provider will be reachable through a dedicated gateway, plus a dedicated airport management network where another AAA server will be deployed.

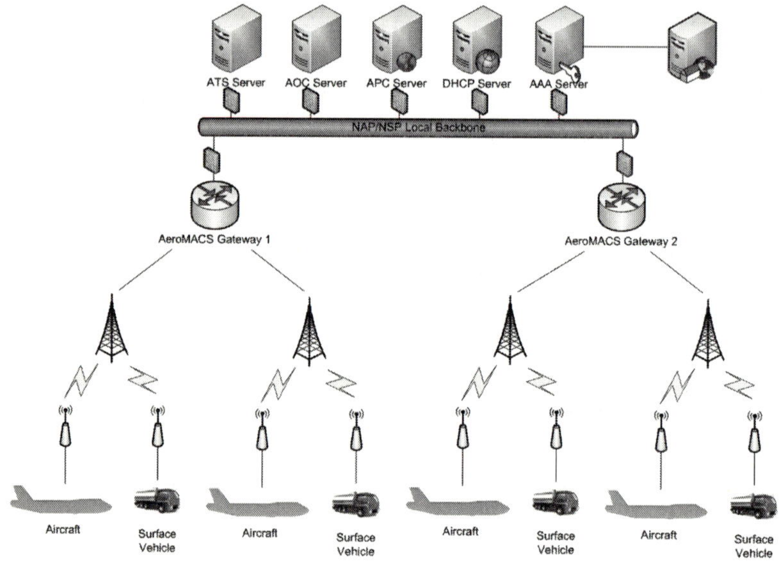

Figure 4.7. *Integrated AeroMACS network topology*

Figure 4.7 illustrates the end-to-end AeroMACS network topology used in the SESAR project. Four main segments can be identified in the topology: the AeroMACS ASN that includes the AeroMACs components and all the functionalities required to establish a radio link in a cellular network. The AeroMACS core is, in this case, composed of the components necessary to provide a local end-to-end connection, which are the NAP AAA server and the Home Agent (HA). The border router separates this domain operated by the airport staff from the domain operated by an external provider. The NSP domain is operated by an external service provider. For registrations managed by the NSP, the local AAA server acts as a proxy that redirects the requests and responses to the NSP AAA server. Finally, the aircraft network is the last segment considered in the global architecture. These two different topologies (isolated and integrated) allow us to consider different levels of complexity for risk assessment. Detailed experimental results will be introduced in section 5.2 and for each scenario, security guidance will be drawn.

5

Aeronautical Network Case Study

This last chapter specifically focuses on the network risk study which is conducted for several scenarios. Thus, a comparison between AeroMACS authentication/authorization security mechanisms is achieved and some security guidance are given either to enhance security policies or to improve the end-to-end security using some additional mechanisms. This case study illustrates how the risk assessment methodology presented in Chapter 3 can help network designers to improve network topologies and re-enforce the global network security level.

5.1. Experimental parameters

5.1.1. *Testbed infrastructure*

This section introduces the testbed platform used to validate the risk assessment algorithm detailed in Chapter 3. The emulation MARIONNET tool [LOD 08] has been used to run different virtual machines for each network node. Figure 5.1 illustrates the emulated risk assessment framework.

MARIONNET allows users to configure and emulate complex networks. The vulnerability scanning tool NESSUS has been installed and configured in a client mode on each node. A NESSUS daemon (the server part) has been also configured on the physical terminal (basically, this is the host where the entire software framework is installed) and used to emulate the network. When the risk assessment algorithm starts, the NESSUS daemon scans every node to get the set of individual vulnerabilities that are detected (see step 2 of the risk assessment process). As long as NESSUS is connected with the NVD vulnerability database, such a platform has the advantage to detect new vulnerabilities if some are newly added to the database and to provide an automated risk assessment tool.

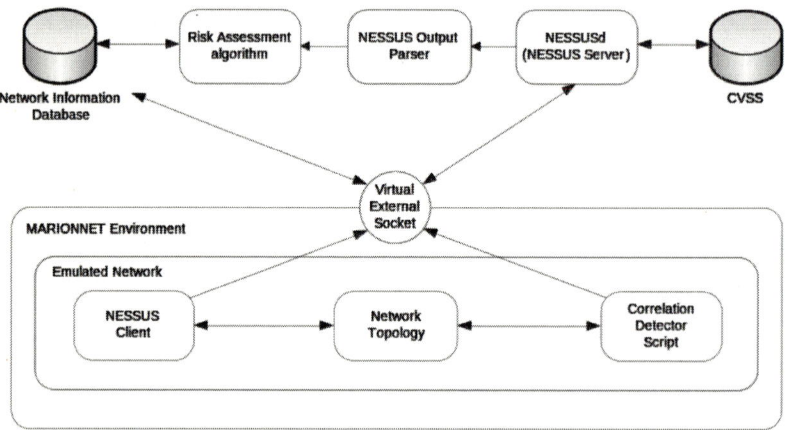

Figure 5.1. *Risk assessment framework using MARIONNET*

The communications between the client and the server parts are made possible due to a special virtualized item in MARIONNET called the Virtual External Socket. The external socket device provides opportunities to connect virtual nodes to non-virtual networks (Internet, for instance). This device allows us to transfer information from the emulated network to the physical terminal. As we used this risk assessment framework for aeronautical network purposes, each aeronautical class (ATS and AOC) is identified using a different network IP address pool. A correlation detector script has been written and distributed over the network to gather correlation information from each node. NESSUS outputs are parsed and all collected network data are then stored in a Network Information Database (IP addresses, CVSS scores, correlation information, etc). Finally, the risk assessment algorithm accesses this database to compute and display information on the network risk level.

5.1.2. *Aeronautical node values instantiation*

In order to instantiate the $ClassValue_i$ and $FunctionValue_i$ parameters (see formula [3.6] for details) for the aeronautical context, the following decomposition has been considered:

– five function nodes with their respective values as shown in Table 5.1.2:

	Firewall gateway	Router	Switch hub	Server	Terminal
Function value	1.0	0.7	0.5	0.3	0.1

Table 5.1. *Node function values*

– In order to be compliant with the ICAO SARPs[1] the following class values are proposed for the two different types of traffic that will be considered in this study case:

	ATS	AOC
Class value	1.0	0.7

Table 5.2. *Aeronautical class values*

The matrix resulting from the combination of these function and class values is:

1 Currently, ATS communications have to be strictly separated from other types, of communications, because of safety and regulatory reasons, as required by ICAO SARPs (Standards and Recommended Practices) [ICA 05]. ICAO SARPs define a set of end-to-end protocols and operational access procedures that permit both safety and non-safety aeronautical applications to use datalink technologies.

	ATS	AOC
Firewall gateway	1.0	0.7
Router	0.7	0.49
Switch hub	0.5	0.35
Server	0.3	0.21
Terminal	0.1	0.07

Table 5.3. *Node values for aeronautical network domains*

Note that the indicated values in Table 5.3 do not integrate the number of connected nodes (n_i) as required in formula [3.6] because this parameter depends on the considered network topology. However, the specific n_i in the value node computation will be integrated depending on the network topology scenario that is considered for testing.

5.1.3. *Aeronautical services instantiation*

The *Security Objective Vector* SOV_s is deduced from the COCR document where security objectives are expressed for three security services (confidentiality, integrity and availability) using a qualitative scale. In order to compute the network risk and get quantitative values, the following mapping has been performed:

Qualitative values	Quantitative values
None	0
Low	1
Medium	2
High	3
High-severe	4
High-catastrophic	5

Table 5.4. *From qualitative to quantitative COCR values mapping*

For instance, if the service s is the ATS data-based Flight Plan Consistency (FLIPCY) service, the corresponding *Security Objective Vector* will be equal to $SOV_{FLIPCY} = [1\ \ 4\ \ 2]$ as indicated by the COCR document.

5.1.4. *Isolated vs. end-to-end emulation scenarios*

Two seperated scenarios have been considered. The first, AeroMACS isolated topology, takes into account only AeroMACS security assumptions. There are no additional network security mechanisms that can interfere with AeroMACS risk assessment. This scenario is important to get a reference risk result in order to provide security guidance for AeroMACS. When these reference results are achieved, a more realistic network topology can be considered. Then, an end-to-end topology is introduced where AeroMACS is no longer the only segment of the global communication system. This scenario is more complex but provides interesting security guidance, in particular additional network agents (such as HA and firewalls) are throughly investigated. Those two scenarios are equivalent to the two different topologies (isolated and integrated) that have been depicted in section 4.3.4.1.

In section 5.2, the behavior of the risk assessment methodology is discussed regarding the network topologies relevant to the AeroMACS system.

5.2. AeroMACS case study: experimental results

5.2.1. *Main inputs for emulation scenarios*

Table 5.5 summarizes the main experimental inputs used for the AeroMACS network scenarios that will be analyzed in the following sections.

5.2.2. *Isolated AeroMACS scenario: preliminary results*

5.2.2.1. *Individual risks*

Nine base stations, 10 aircraft and 12 surface vehicles have been considered in this simulation. Figure 5.2 illustrates the individual risks assessed for each node in the network.

Node ID	Connected nodes	Security protection	Vulnerabilities
Base stations (1–6)	3	8	1
Base stations (7 and 8)	4	8	1
Base station 9	5	8	1
Aircraft	1	2	0
Surface vehicles	1	2	0
AAA server	1	3	20
DHCP server	1	1	64
ASN gateway	14	2	1
ATS server	1	1	47
AOC server	1	1	47
APC server	1	1	13

Table 5.5. *AeroMACS topology emulation parameters*

Base stations and ASN gateway individual risks are relatively low because there is only a single vulnerability detected for these nodes. Despite having the same specific vulnerability (CVE-2008-1542)[2], there is a difference (evaluated from 0.075 to 0.15) between base station 8, base station 9 and the seven first base stations' individual risks. This is mainly due to the higher number of connected nodes of base station 8 and base station 9 which increases their node values (all the base stations are not connected to the same nodes as described in Table 5.5).

Another interesting fact is that the APC server and the ASN gateway's individual risks are nearly close (respectively 0.46 and 0.5) despite a big difference in the number of intrinsic vulnerabilities on each of them (respectively 13 and 1). Indeed, a higher individual risk will be expected for the APC server node as long as it has more

2 Vulnerability details as it appears in the NVD: base station has "topsecret" as its password for the root account, which allows remote attackers to obtain administrative access via a telnet login.

vulnerabilities, however, the ASN gateway compensates the gap by a higher node value in the network (equal to 14) whereas the APC server, giving it functionality and traffic class value, is the lowest one (equal to 0.03).

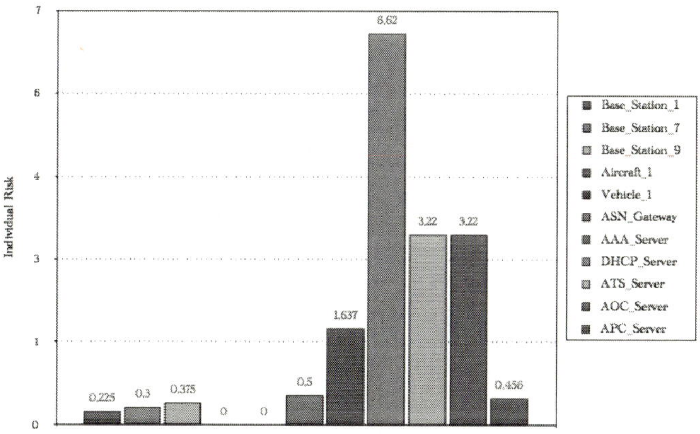

Figure 5.2. *Individual risks for all network nodes*

The DHCP server node is the most vulnerable node in the network, and consequently has the highest individual risk out there (assessed to 6.62). Sixty-four vulnerabilities have been detected for that node with very high CVSS scores: 92.12% of these have top CVSS score (meaning 10, the highest score in the NVD database). Even the lowest CVSS score is relatively high (9.3) if it is compared to base stations or ASN gateway vulnerability scores (respectively 7.5 and 5.0). Finally, the AAA, ATS and AOC servers, regarding the assumptions made in the inputs, get medium individual risk values.

Except the ASN gateway individual risk value that is considerably impacted by the high node connectivity, all the individual risk values that have been measured seem to increase with the number of exploitable vulnerabilities per node taken from the NVD database as shown in Figure 5.3.

Regarding the assumptions made for the inputs of the current simulation, it can be deduced that the individual risk depends on the number of vulnerabilities per nodes.

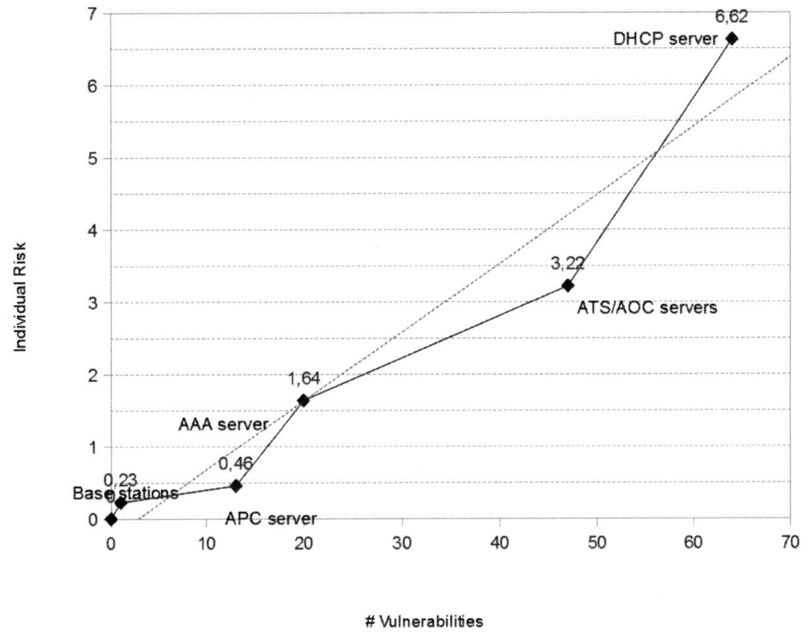

Figure 5.3. *Individual risk evolution as a function of vulnerabilities for all nodes*

Table 5.6 illustrates the repartition of the CVSS score for each node in the network (the same terminology and taxonomy used in the NVD database has been respected):

CVSS score	Number of vulnerabilities	Percentage %
$[0,1]$	0	0
$[1,2]$	2	0.995
$[2,3]$	1	0.498
$[3,4]$	0	0
$[4,5]$	7	3.483
$[5,6]$	11	5.473
$[6,7]$	4	1.990
$[7,8]$	111	55.224
$[8,9]$	0	0
$[9,10]$	65	32.338
Total	**201**	–
Average score	**7.938**	

Table 5.6. *Vulnerability CVSS statistics*

The majority of the scores are ranked in the [7, 8] NVD CVSS interval (and represent 55.224% of the total vulnerability scores). The maximum CVSS scores ranked between 9 and 10 are in most cases relative to the DHCP server node that explains why this node has the highest individual risk value among the network. The average CVSS score has been measured to 7.938.

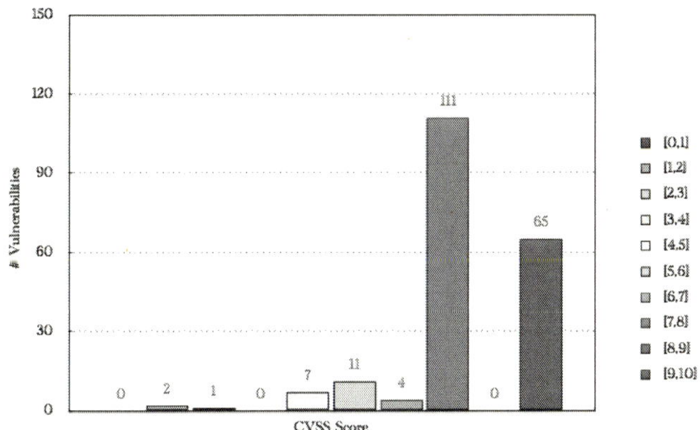

Figure 5.4. *Vulnerability CVSS statistics*

5.2.2.2. *Propagated risks*

Table 5.7 summarizes the propagated risk values for each node in the network. The propagated risk results are mainly impacted by the importance of the connected node number parameter in the algorithm. For instance, some assumptions have been made regarding the topology of this scenario for the base stations: the first six base stations are connected to three nodes (i.e. one aircraft, one vehicle and the ASN gateway), base stations 7 and 8 are connected to four nodes (plus another vehicle) and the last base station to five nodes. The others parameters (security protection, offered service, exchanged data and NVD vulnerabilities) remain the sames. However, the propagated risk values are slightly different (ranging from 7.474 to 12.456) because of different correlation density in the network.

Node ID	Propagated risk
Base stations (1–6)	7.474
Base stations 7 and 8	9.965
Base station 9	12.456
Aircraft (1–6)	0.812
Aircraft (7–12)	1.082
Vehicle (1–6)	0.812
Vehicle 7 and 8	1.082
Vehicle 9 and 10	1.353
ASN gateway	538.998
DHCP and AAA servers	1.2
ATS server	0.398
AOC server	0.750
APC server	0.135

Table 5.7. *Propagated risks for all network nodes*

The propagated risk for the aircraft also deserves to be deeply discussed. Indeed, it is not the same for the first six aircraft (equal to 0.812) as the last six aircraft (equal to 1.082). However, the justification does not lie in the connected node parameter this time as far as all aircraft are connected to a single base station. The difference between propagated risk for the aircraft (assessed to 0.27) is due to the individual risk specific to the base station to which the aircraft is connected to.

As Figure 5.2 depicts, the individual risk values can be different from one base station to another because of the node repartition decided in the network topology. DHCP, AAA, ATS, AOC and APC servers have all low propagated risk values (ranging from 0.135 to 1.2) because all of them are connected to a single node (the ASN gateway) that has a very low individual risk (equal to 0.5).

Moreover, the most important result in this scenario is the propagated risk value of the ASN gateway that supersedes all the remaining nodes. This is likely due to a high node correlation of the ASN gateway. As far as it is the "corner stone" of the topology where all node exchanges have to pass through the gateway, it is logically impacted by the other nodes. The concept of propagated risk lies in the importance of the connected node number of parameter as depicted in Figure 5.5.

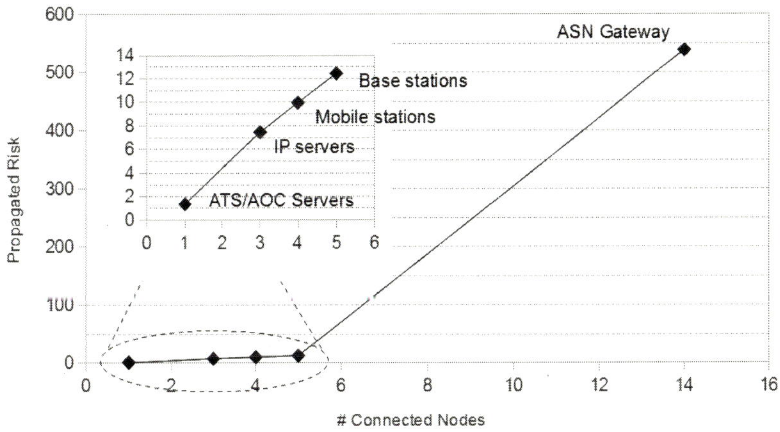

Figure 5.5. *Propagated risk evolution as function of connected nodes for all network nodes*

5.2.2.3. *Node and network risks*

The node risk of the ASN gateway is hardly impacted by the high propagated risk of the node (see formula [3.1] of node risk evaluation). Furthermore, the high node value of the ASN gateway plays a major role in the growth of the node risk value: even the highest node risk (which is relevant to the base station 9) is 117 times smaller than the ASN gateway node risk.

Consequently, as the network risk is given by the sum of the node risk, the network risk is mainly represented by the ASN gateway node risk (see Figure 5.6 for details). The detailed node risk repartition compared to the network risk is given in Table 5.8.

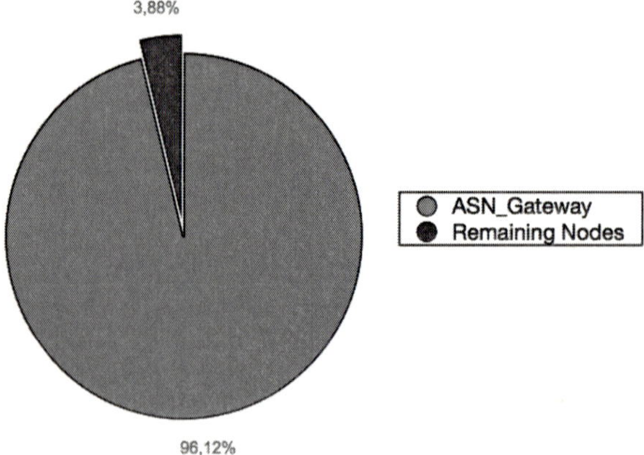

Figure 5.6. *Percentage of network risk per node risk*

Figure 5.7 shows the contribution (%) of each connected node with the ASN gateway propagated risk. It is clear that the DHCP server represents the biggest contribution in the propagated risk of the ASN gateway node and is the main actor of this high risk value.

This means that if the final goal is to get the ASN gateway node risk lower (and consequently the network risk), it is mandatory to tweak two related parameters:

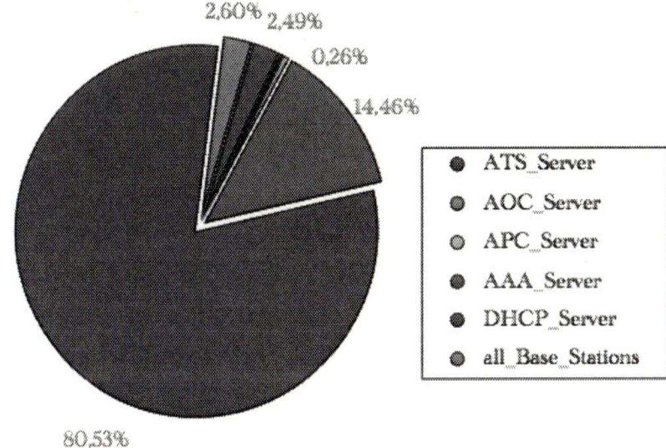

Figure 5.7. *ASN gateway propagated risk contribution per connected node*

1) The number of connected nodes is important. It should be limited for each node as much as possible. This can be done by some topological considerations that allow the network risk to be less higher than in this first simulation.

2) The ASN gateway is clearly the point of failure of this topology (see the previous results). Despites its high node connectivity, its high propagated risk value is not directly due to the number of interconnected nodes, but rather to the high individual DHCP server individual risk, the high correlation that exists between the two nodes (the ASN gateway is the only node connected with the DHCP server node) and the high node value of the ASN gateway.

Node ID	Node risk	Network risk (%)
All base stations	284.866	3.63
All aircraft	8.337	0.11
All vehicle	6.821	0.09
ASN gateway	7552.98	96.12
DHCP server	2.346	0.03
AAA server	0.851	0.01
ATS server	1.085	0.01
AOC server	0.833	0.01
APC server	0.017	0

Table 5.8. *Node risk results*

5.2.3. *Isolated AeroMACS scenario: EAP vs. RSA sub-scenario*

As has been previously mentioned, the AeroMACS privacy sub-layer is able to support both EAP and RSA for authentication and authorization security services. The aim of this scenario is to compare the effects of these security options on the AeroMACS network risk. Currently, only one option can be chosen, which is using RSA or EAP as an authentication and authorization protocol.

Many vulnerabilities have been found for both security mechanisms: the NVD database clearly indicates a higher number of vulnerabilities for RSA (i.e. 33 vulnerabilities) compared to EAP (only 4 vulnerabilities). Indeed, RSA is much better known and used over all IT systems around the world. Thus, it is quite logical to find more vulnerability inputs in the database compared to EAP.

Moreover, the vulnerability statistics made in this scenario show again that the number of vulnerabilities is not the only indicator to be taken into account on which security mechanism should be favored. Indeed, despite a higher number of vulnerabilities (+29 vulnerabilities), RSA remains more secure than EAP: as shown in the previous simulation, the average CVSS score is a safer criterion if the goal is to compare two or more security mechanisms in the NVD database. The average CVSS score (on the total vulnerabilities in the AeroMACS network) has been evaluated to 6.325 for RSA and 7.795 for EAP. Table 5.9 provides more details on the vulnerabilities statistics for the EAP versus RSA scenario simulations:

However, the average CVSS score should be weighted according to the individual risk values obtained after the simulation. Indeed, Figure 5.9 shows the individual risk values updated for the base stations and the ASN gateway (all the remaining nodes are not represented since there is no noticeable change).

CVSS score	EAP Vulnerabilities	EAP Percentage (%)	RSA Vulnerabilities	RSA Percentage (%)
[0, 1]	0	0	0	0
[1, 2]	2	0.83	12	2.26
[2, 3]	1	0.415	11	2.072
[3, 4]	0	0	10	1.883
[4, 5]	7	2.905	107	20.151
[5, 6]	31	12.863	121	22.787
[6, 7]	4	1.66	24	4.52
[7, 8]	121	50.207	161	30.32
[8, 9]	0	0	0	0
[9, 10]	75	31.12	85	16.008
Total	**241**	–	**531**	–
Average score	**7.795**		**6.325**	

Table 5.9. *EAP vs. RSA vulnerability statistics (all nodes)*

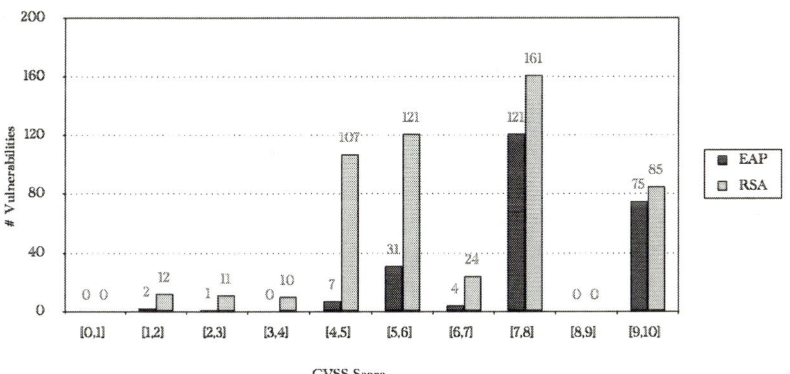

Figure 5.8. *Vulnerability CVSS score distribution for EAP and RSA*

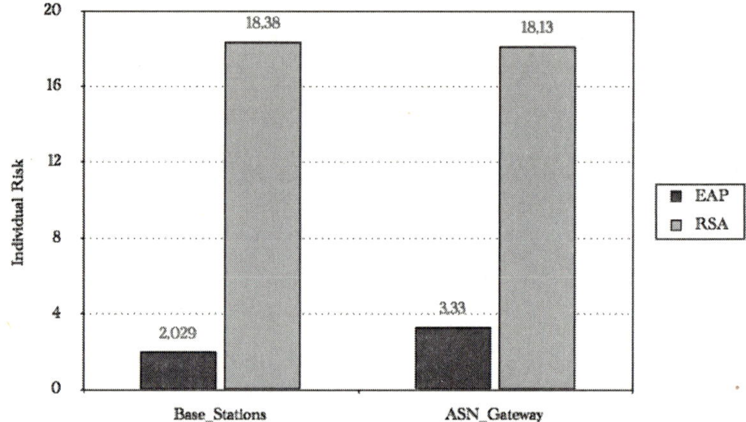

Figure 5.9. *Individual risks for base stations and the ASN gateway*

The higher number of vulnerabilities for RSA naturally makes the individual risk higher than EAP for both base stations and the ASN gateway (+16.35 and +14.8, respectively, for RSA and EAP). These results suggest first that the number of vulnerabilities remains an important parameter because the individual risk is computed as the sum of the likelihood of occurrence of a threat and its impact on the total number of vulnerabilities: since RSA has much more inputs in the NVD database, the individual risk relevant to EAP is lower.

In conclusion, to consider the node risk individually, it is clear that EAP should be used for authentication and authorization in the AeroMACS nodes.

However, the propagated risk results should also be considered to effectively construct the final guidance on the use of a EAP or RSA protocol. Figure 5.10 shows the propagated risk values using a EAP or RSA protocol for all network nodes (except the ASN gateway that has a very big propagated risk value and is not represented for clarity matters). The same comments for the individual risks remain true here: the EAP authentication protocol induces a lower propagated risk compared to the RSA protocol.

The ASN gateway is still the point of failure in both sub-scenarios since it has the largest propagated risk among all the network nodes (1042.64 and 2499.87, respectively, for EAP and RSA). It still has the biggest contribution in the global network risk (either for EAP or RSA) as illustrated in Table 5.10.

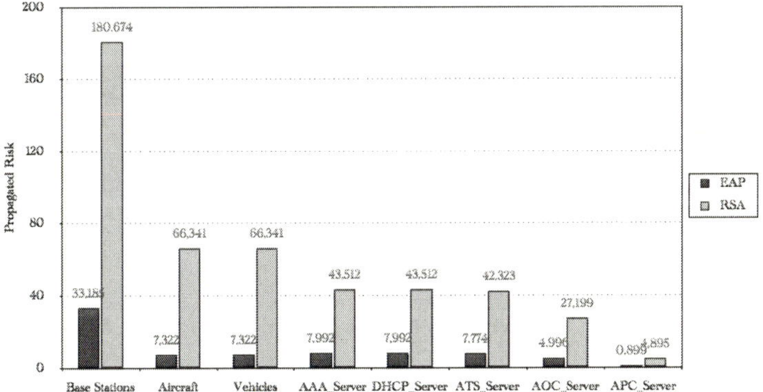

Figure 5.10. *Propagated risks for all nodes (EAP vs. RSA)*

Node ID	Network risk (%)	
	EAP	RSA
All base stations	6.867	14.509
All aircraft	0.386	1.310
All vehicle	0.322	1.091
ASN gateway	92.118	82.888
DHCP server	0.03	0.038
AAA server	0.053	0.030
ATS server	0.135	0.0755
AOC server	0.081	0.0546
APC server	0	0

Table 5.10. *Node risk statistics – EAP vs. RSA*

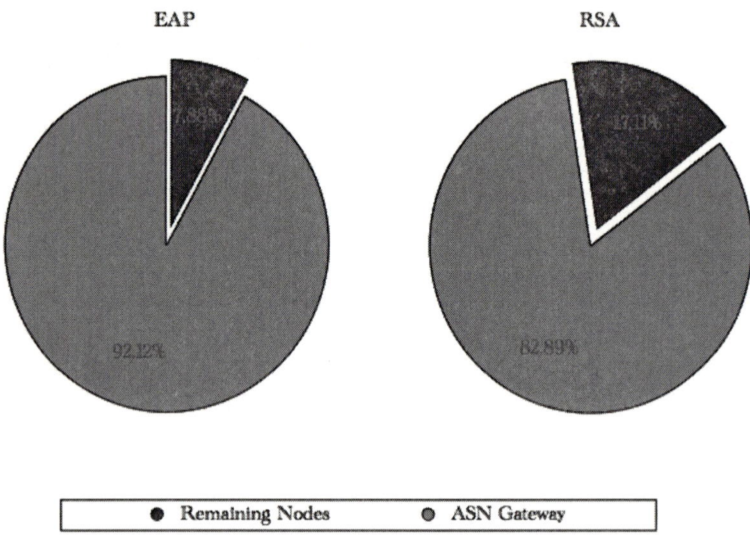

Figure 5.11. *Percentage of network risk per node risk (EAP vs. RSA)*

Regarding the experimental results presented above, the big lines of the preliminary guidance that should help to decrease the risk level for the different network nodes as depicted in the next subsection can already be drawn.

5.2.4. *Preliminary AeroMACS security enhancement guidance*

Based on the previous experimental results the following preliminary security guidance can be proposed. Each of them should be able (see in the following scenario for details) to enhance the AeroMACS network security. Those guidances are divided into three categories:

1) *Implementation guidances*: network nodes should be chosen wisely with a minimum of intrinsic vulnerabilities. IP Commercial off-the-Shelf (COTS) nodes (AAA and DHCP servers) should be discussed regarding the number of the exploitable vulnerabilities and their respective CVSS scores. It could be interesting to establish a state of the art of the potentially usable IP nodes (particularly the DHCP server node), classify them by number of vulnerabilities and

CVSS scores and see how the individual risk per node is affected. The preferred nodes are obviously those with the lowest individual risks.

2) *Topological guidances*: the global network risk is highly impacted by the propagated risk values (more than the individual risk values) because the node connectivity is taken into account at this step of the risk assessment process. It is clear that the ASN gateway is the main issue in this topology and some countermeasures should be taken to avoid this problem. For instance, according to the AeroMACS network reference architecture depicted in Figure 4.5, it could be interesting to select two ASN gateways as NAPs, each one connected to a set of base stations and IP nodes. This will likely provide less node correlation between the ASN gateway (then a lower likelihood of correlation) and highly impacted IP nodes (such as the DHCP server) and consequently decreases the network risk.

3) *AeroMACS security guidances*: the most constraining node in the network and their respective contribution in the global network risk are clearly identified. Thus, some security mechanisms should be deployed to limit the propagated risk. Particular attention should be given to the connectivity between the ASN gateway and the IP nodes such as the DHCP server. To solve this connectivity problem, firewalls should be privileged. Indeed, they can limit the data exchanges between a highly vulnerable node and the ASN gateway. Also, maximizing security protections at a layer 2 (typically AeroMACS security) should help the propagated node risk decrease for AeroMACS-based nodes (i.e. base stations and mobile stations). In this context, the EAP versus RSA simulations showed that EAP induces a lower risk (individual, propagated and network risk) for the isolated AeroMACS topology.

5.2.5. *AeroMACS implementation improvements: isolated scenario without operational server vulnerabilities*

After the preliminary guidance provided by the first experiment campaign, the isolated scenario has been extended. The first idea is to start the experiments again with the same parameters except for the operational aeronautical servers (i.e. ATS, AOC and APC servers), for which the intrinsic vulnerabilities have been removed (because they relied on COTS products, which has not been realistically considered). The second idea is to take into consideration the topological guidances, using two ASN gateways instead of one, in order to reduce the propagated risk between the network nodes.

The results presented in this section are (in most cases) compared to the first experimental results with the aim of discussing whether the new assumptions are helpful or not to improve the security level of the whole network architecture. This closed control loop process is an example of how our methodology can help network designers to improve the network security of their IT organization.

5.2.5.1. *Experimental inputs*

Table 5.11 summarizes the main emulation inputs used in this study. Parameters are kept unchanged except for the number of vulnerabilities parameters relevant to the operational servers.

Node ID	Function value	Class value	Number of connected nodes	Number of security protection	Number of vulnerabilities
Base stations	1	1	3/4/5	8	1
Aircraft	0.7	1	1	2	0
Surface vehicles	0.7	1	1	2	0
AAA server	0.3	1	1	3	20
DHCP server	0.3	1	1	1	64
ASN gateway	0.3	1	14	2	1
ATS server	0.3	1	1	1	0
AOC server	0.3	0.7	1	1	0
APC server	0.3	0.1	1	1	0

Table 5.11. *Emulation parameters (without operational server vulnerabilities)*

5.2.5.2. *Network topology*

The network topology is depicted in Figure 5.12 (which is by the way the same topology chosen in the previous experiments). Basically, it remains the same as there is no added or removed node in the network.

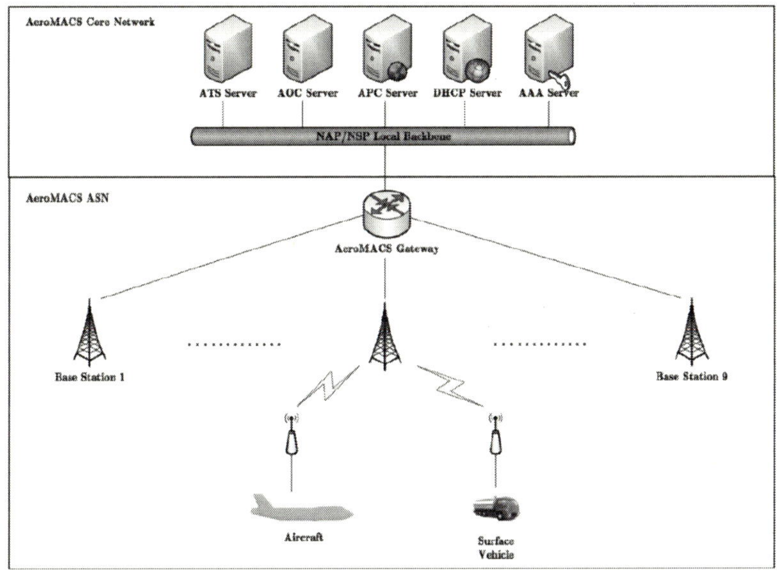

Figure 5.12. *AeroMACS network topology: extended isolated scenario*

5.2.5.3. *Vulnerability statistics*

Table 5.12 shows the updated vulnerability CVSS statistics after removing the operational aeronautical server vulnerabilities.

The most interesting result here is probably the fact that even if the total number of vulnerabilities is reduced (from 201 to 94), the average CVSS score is higher in this extended scenario (8.859 against 7.938 as depicted in Table 5.12). This result is mainly due to the new distribution of CVSS scores that put a higher emphasis on the critical vulnerabilities (i.e. [9, 10] CVSS score range) in this scenario compared to the previous scenario. This obviously has an impact on the individual risk results, but should not alter the propagated risk values as this type of risk is more sensitive to the node correlation parameter. Figure 5.13 shows the distribution of CVSS scores in each scenario.

CVSS score	Number of total vulnerabilities	Percentage (%)
[0, 1]	0	0
[1, 2]	0	0
[2, 3]	0	0
[3, 4]	0	0
[4, 5]	2	2.127
[5, 6]	9	9.574
[6, 7]	2	2.127
[7, 8]	16	17.021
[8, 9]	0	0
[9, 10]	65	69.148
Total	**94**	–
Average score	**8.859**	

Table 5.12. *Updated vulnerability CVSS statistics*

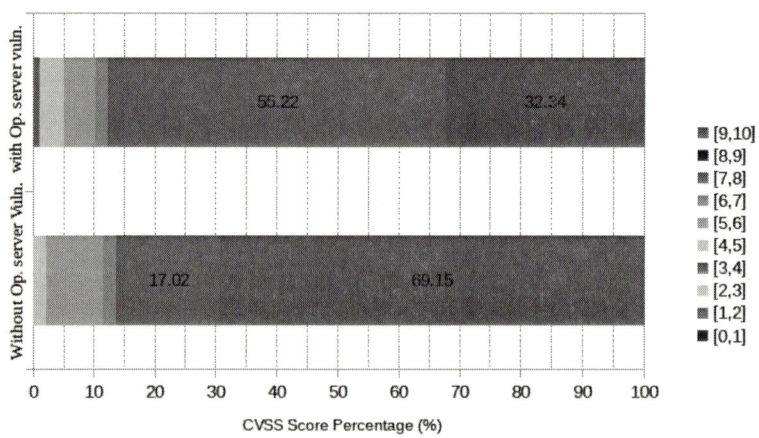

Figure 5.13. *Comparison of CVSS score distribution (with and without operational server vulnerabilities)*

5.2.5.4. *Individual risk results*

Except for the operational servers, individual risk results are not altered. Indeed, beside the number of connected nodes (which is taken into account in the node value parameter), there is no consideration for node correlation in the individual risk. That being said, removing the operational server vulnerabilities does not have any impact on the other individual risk results relevant to the remaining network nodes (even those directly connected to them such as the ASN gateway).

Table 5.13 summarizes the individual risk values for each node.

Node ID	Individual risk
Base stations (1–6)	0.23
Base stations 7 and 8	0.3
Base station 9	0.38
Aircraft (all)	0
Vehicle (all)	0
ASN gateway	0.5
DHCP server	6.62
AAA server	1.64
ATS server	0
AOC server	0
APC server	0

Table 5.13. *Updated individual risk results*

5.2.5.5. *Propagated risk results*

As shown in Table 5.14, removing the operational server vulnerabilities undeniably has an impact on the propagated risk values for all nodes in the network. Updated propagated risk values have decreased compared to values in Table 5.7.

Node	Propagated risk
Base stations (1–6)	5.471
Base stations 7 and 8	8.012
Base station 9	9.586
Aircraft (1–6)	0.355
Aircraft (7–12)	0.941
Vehicle (1–6)	0.355
Vehicles 7 and 8	0.670
Vehicles 9 and 10	0.941
ASN gateway	417.013
DHCP and AAA server	0.66
ATS server	0.182
AOC server	0.551
APC server	0.054

Table 5.14. *Updated propagated risk results*

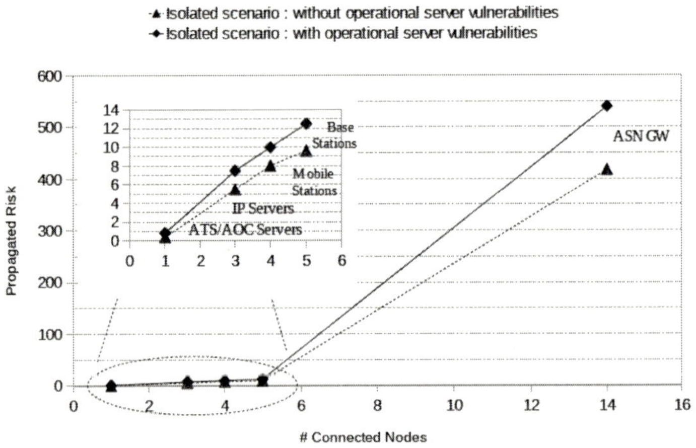

Figure 5.14. *Comparison of propagated risks as a function of the number of connected nodes (with and without operational server vulnerabilities)*

This difference is clearer when the propagated risk evolution is compared as a function of the number of connected nodes for both scenarios as depicted in Figure 5.14: for all network nodes, propagated risk values are lower and decrease in the new scenario (i.e. without operational server vulnerabilities). This is an important result as it shows that even if some assumptions have been made regarding a node individually (for instance, by updating its exploitable vulnerabilities or CVSS scores), it has an impact on the propagated risk values relevant to logically connected nodes (i.e. there is a data flow exchange between them) to that specific node.

5.2.5.6. *Network risk results*

The network risk results are impacted in the same way as the propagated risk results, the new values are summarized in Table 5.15.

Node	Node risk	Network risk (%)
All base stations	218.838	3.601
All aircraft	4.687	0.077
All vehicles	3.748	0.061
AAA server	0.689	0.011
DHCP server	2.184	0.03
ATS server	0.054	0
AOC server	0.115	0.001
APC server	0.001	0
ASN gateway	5845.192	96.209

Table 5.15. *Updated network and node risk results*

Despite a decrease of every node risk value, the ASN gateway has always the biggest node risk among all nodes and its contribution in the network risk always overseeds the other nodes. Figure 5.15 shows a comparison between node contributions in the network risk for both scenarios.

In conclusion, it can be said that removing the operational server vulnerabilities has an impact on both individual risk (relevant to ATS, AOC and APC servers) and propagated risk values but the network risk value remains high because the node correlation was not modified, and then the ASN gateway is still the Achille's heel of the isolated topology. To address this issue, in the following section, another experiment campaign is described using two ASN gateways, instead of one, and by keeping the assumptions made in this section for the operational server vulnerabilities.

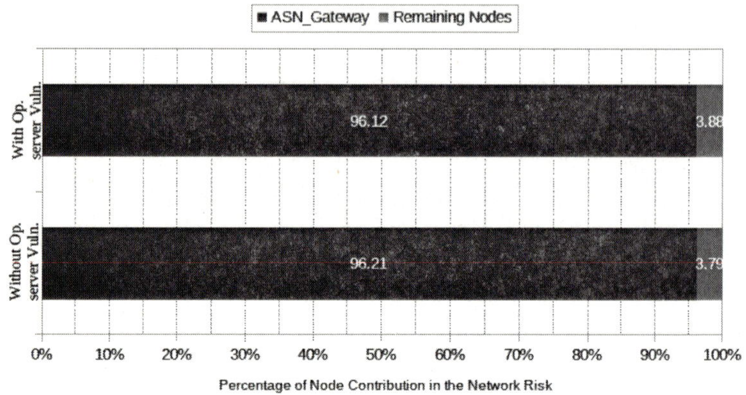

Figure 5.15. *Comparison of the percentage of network risk per node risk (with and without operational server vulnerabilities)*

5.2.6. *AeroMACS topological improvements: isolated scenario with two ASN gateways*

5.2.6.1. *Experimental inputs*

Table 5.16 summarizes the main emulation inputs that are used in this study. The impact of the new topology resulting from the integration of a second ASN gateway implies the following changes:

– Servers behind the backbone are now connected to two ASN gateways (instead of only one).

– The node correlation for ASN gateways is now reduced to 9 (or 10 depending on the connected BSs) instead of 14 in the first simulation.

– The assumptions made for operational servers in the previous simulation (i.e. no COTS vulnerabilities) are maintained.

Node ID	Function value	Class value	Number of connected nodes	Number of security protection	Number of vulnerabilities
Base stations	1	1	3/4/5	8	1
Aircraft	0.7	1	1	2	0
Surface vehicles	0.7	1	1	2	0
AAA server	0.3	1	2	3	20
DHCP server	0.3	1	2	1	64
ASN gateway 1	0.3	1	9	2	1
ASN gateway 2	0.3	1	10	2	1
ATS server	0.3	1	2	1	0
AOC server	0.3	0.7	2	1	0
APC server	0.3	0.1	2	1	0

Table 5.16. *Simulation parameters (scenario with two ASN gateways)*

5.2.6.2. *Network topology*

The network topology has changed with the addition of the second ASN gateway as depicted in Figure 5.16.

This topological amendment will likely reduce the correlation between a single ASN gateway and connected nodes, as it has been considered as the central hub and point of failure in the first topology. With the introduction of a second ASN gateway and removing the operational server vulnerabilities, the risk values should decrease.

5.2.6.3. *Vulnerability statistics*

The vulnerability statistics are the same as the statistics presented in Table 5.12.

5.2.6.4. *Individual risk results*

As for the individual risk results discussed previously, the altered values are relevant to nodes that have been impacted by the new topology. Then, the individual risk results remain the same for all the nodes except the IP servers (i.e. AAA and DHCP servers) and both ASN gateways.

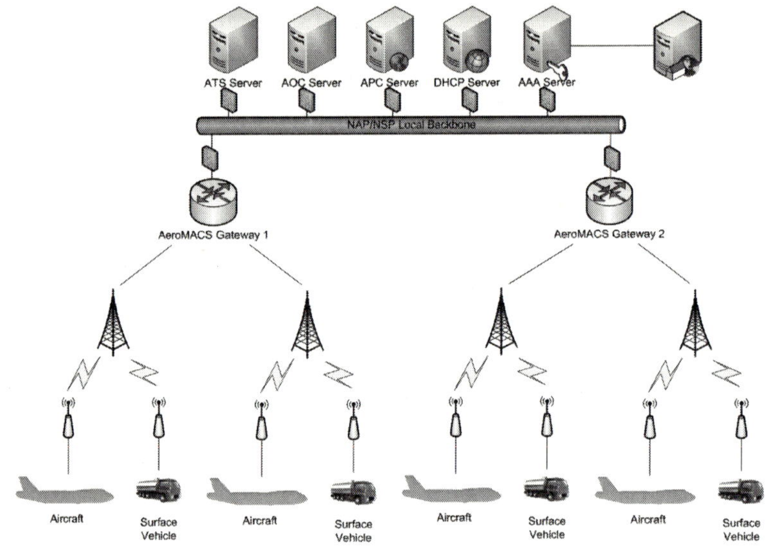

Figure 5.16. *The new isolated network topology using two ASN gateways*

Considering Table 5.18, the ASN gateway risk values are lower compared to the scenario with one gateway: indeed, the vulnerabilities (and their respective scores) are the same whereas the node connectivity has decreased (from 14 to either 9 or 10). The difference between the individual risk values for the two ASN gateways is due to the number of connected nodes.

Moreover, it has been shown that IP server's individual risk values have been doubled compared to the scenario with just one ASN gateway. Indeed, the DHCP and AAA servers where connected to one node (which is the ASN gateway), whereas in the second scenario they are connected to two nodes (both ASN gateways). As all the parameters relevant to IP servers have not changed except the node connectivity n_i (see formula [3.5] for details), which is taken into account in the likelihood of occurrence of a threat formula (see the first deliverable for more details), the individual risks have been multiplied by a factor of two.

Node ID	Individual risk
Base stations (1–6)	0.23
Base stations 7 and 8	0.3
Base station 9	0.38
Aircraft (all)	0
Vehicle (all)	0
ASN gateway 1	0.32
ASN gateway 2	0.29
DHCP server	13.24
AAA server	3.26
ATS server	0
AOC server	0
APC server	0

Table 5.17. *Updated individual risk values (isolated scenario with two ASN gateways)*

Quantitatively, these new individual risk values show that overall the risk is getting higher: as a matter of fact, adding an additional node induces has increased the intrinsic risk for IP servers. But conclusions should be drawn after analyzing the propagated risk results, and more importantly the overall network risk. Hence, these results should be confronted to the propagated risk values (and later the node risk values) presented in the following section.

5.2.6.5. *Propagation risk results*

As expected, adding another ASN gateway makes the propagated risk values increase or decrease, depending on the node correlation in the new topology as shown in Table 5.18.

Unlike the previous simulation scenarios, base stations 1–6 do not have the same propagated risk values (see Table 5.18). Indeed, base stations 5 and 6 are connected to ASN gateway 2 (with a higher node correlation), which explains the higher propagated risk values for these base stations compared to base stations 1–4 (connected to the first ASN gateway 1). Propagated risk values for base stations 7–9 have also decreased.

Node	Propagation risk
Base stations (1–4)	4.469
Base stations 5 and 6	4.589
Base stations 7 and 8	6.062
Base station 9	7.518
Aircraft (1–4)	0.316
Aircraft 5 and 6	0.329
Aircraft (7–12)	0.549
Vehicle (1–6)	0.754
Vehicles 7 and 8	0.549
Vehicles 9 and 10	0.754
ASN gateway 1	268.21
ASN gateway 2	260.019
DHCP and AAA server	0.691
ATS server	0.250
AOC server	0.785
APC server	0.093

Table 5.18. *Updated propagated risk values (two ASN gateways)*

The observation is made for the propagated risk values relevant to the aircraft 1–6. In the previous scenarios, as they were connected to the same gateway, they have the same propagated risk values (see also Tables 5.13 and 5.14), but in this case, despites having the same node correlation, aircraft 5 and 6 have slightly different propagated risk values as they are connected to the second gateway.

Note that the DHCP and AAA server propagated risk values are the only results that have increased because of the higher node correlation in this scenario (two instead of one). For the ASN gateways, as the node correlation has decreased (from 14 nodes to 9 and 10) in this scenario compared to the previous scenario, it is quite logical to obtain a lower propagated risk for both ASN gateways.

In summary, propagated risk values relevant to nodes behind the backbone (belonging to the AeroMACS core network) have slightly increased whereas those relevant to nodes in the AeroMACS access service network have significantly decreased. These unbalanced changes in the propagated risk values have an impact on the network risk results as depicted in section 5.2.6.6.

5.2.6.6. *Network risk results*

Table 5.19 summarizes the values for the node risks of all nodes in the new topology.

Node	Node risk
Base stations (1–4)	14.082
Base stations 5 and 6	14.433
Base stations 7 and 8	25.449
Base station 9	39.469
Aircraft (1–4)	0.221
Aircraft 5 and 6	0.230
Aircraft (7–12)	0.384
Vehicle (1–6)	0.221
Vehicles 7 and 8	0.384
Vehicles 9 and 10	0.527
ASN gateway 1	725.031
ASN gateway 2	780.927
DHCP server	8.358
AAA server	2.370
ATS server	0.15
AOC server	0.329
APC server	0.00

Table 5.19. *Updated node risk values (two ASN gateways)*

As expected, the node risk values have substantially decreased compared to those relevant to scenarios using a single gateway, except for nodes in the AeroMACS core network (servers), but their respective contributions in the network risk are not really significant to be emphasized. Figure 5.17 shows a comparison between the network risk values for the three different simulations conducted at this stage of the study.

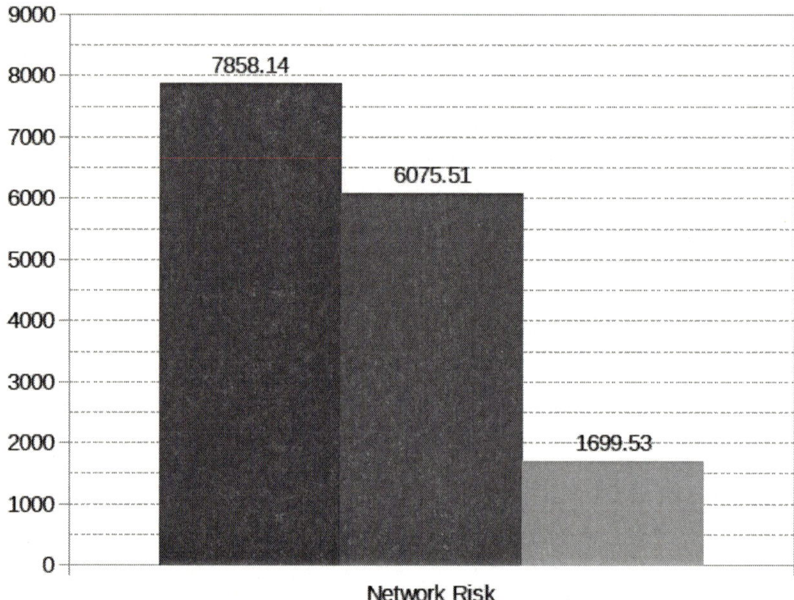

Figure 5.17. *Comparison of network risk values*

It is clear that the isolated scenario using two ASN gateways without operational server vulnerabilities gives better results regarding the risk assessment simulations. The main observations that can be made about this scenario using the new topology with two ASN gateways are the following:

– Individual risk values remained unchanged or decreased except for IP servers because of a higher node value (and then a higher likelihood of occurrence of a threat on these nodes).

– Propagated risk values relevant to nodes at the AeroMACS core network have slightly increased whereas those relevant to nodes at the AeroMACS access service network have considerably decreased.

– The overall network risk value of the new topology is lower compared to the scenario results using a single ASN gateway. This result shows that duplicating the ASN gateways and dispatching as much as possible the base stations among them reduces the propagated risk and consequently the overall network risk.

In the following scenario, a focus is made on the end-to-end topology and on the evolution of the different types of risks.

5.2.7. *Scenario with end-to-end AeroMACS topology*

5.2.7.1. *Experimental inputs*

Table 5.20 summarizes the main experimental inputs used in this integrated topology. With the addition of the firewall and the HA nodes, several parameters have been modified such as the number of connected nodes and security protections.

Node ID	Function value	Class value	Number of connected nodes	Number of security protection	Number of vulnerabilities
Base stations	1	1	3/4/5	8	1
Aircraft	0.7	1	1	2	0
Surface vehicles	0.7	1	1	2	0
AAA server	0.3	1	2	4	20
DHCP server	0.3	1	1	2	64
ASN gateway 1	0.3	1	5	3	1
ASN gateway 2	0.3	1	6	3	1
ATS server	0.3	1	1	2	0
AOC server	0.3	0.7	1	2	0
APC server	0.3	0.1	1	2	0
HA	1	1	2	1	4
Firewalls	1	1	2/3/6	1	1

Table 5.20. *Simulation parameters (Integrated AeroMACS Topology)*

5.2.7.2. *Network topology*

Considering Figure 4.7, different items have been added to the previous network topology (which contains two ASN gateways):

1) A mobile IP HA node that essentially handles the following tasks:
   – registration of mobile stations;
   – routing and forwarding of mobile station traffic.

2) Firewalls to provide protection to:
   – ASN gateways at the entry of the AeroMACS ASN;
   – operational and IP servers;
   – the HA.

Note that with the introduction of the HA to support IP mobility, the data flow exchanges have completely been re-evaluated mainly because of the mobile IP subscriber (could be an aircraft of a surface vehicle) registration process as follows:

1) The ASN gateway sends the subscriber registration request from the mobile station to the mobile IP HA. The sought ASN gateway is obviously the gateway connected to the concerned mobile stations.

2) HA requests that the AAA server (which is now directly connected to the HA node) sends the cryptographic keys for the Mobile IP session. The HA announces to the AAA server that it would like to source IP session-based accounting messages.

3) The AAA server agrees to use IP session-based accounting, provides the requested cryptographic keys and sends the AAA-Session-ID for this session.

4) The HA replies to the mobile IP registration request.

Note that the network topology provided in Figure 4.7 has been provided to match as closely as possible the WiMAX end-to-end network model provided by the WiMAX Forum Network Working Group (NWG) as shown in Figure 5.18:

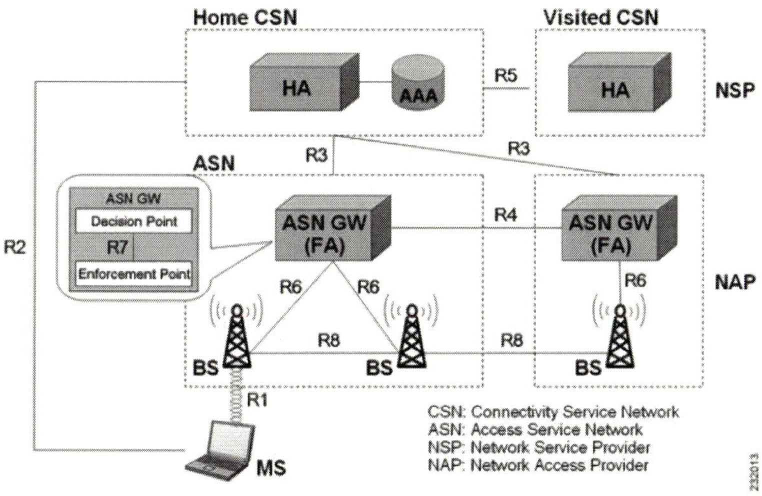

Figure 5.18. *WiMAX forum NWG end-to-end network model*

5.2.7.3. *Vulnerability statistics*

Table 5.21 shows the updated vulnerability CVSS statistics of the integrated network topology.

CVSS score	Number of total vulnerabilities	Percentage (%)
$[0, 1]$	0	0
$[1, 2]$	0	0
$[2, 3]$	0	0
$[3, 4]$	0	0
$[4, 5]$	2	1.980
$[5, 6]$	9	8.910
$[6, 7]$	2	1.980
$[7, 8]$	22	21.782
$[8, 9]$	0	0
$[9, 10]$	66	65.346
Total	94	–
Average score	8.651	

Table 5.21. *Updated vulnerability CVSS statistics (integrated network topology)*

The CVSS distribution is not really very different from the distribution presented in Table 5.21: there are few vulnerabilities added to the final statistics (i.e. one vulnerability per firewall and four for the HA) that made the distribution more balanced, mainly between the highest CVSS score ranges ([8, 9] and [9, 10]). A Comparison between the CVSS score distributions in all simulation scenarios is shown in Figure 5.19.

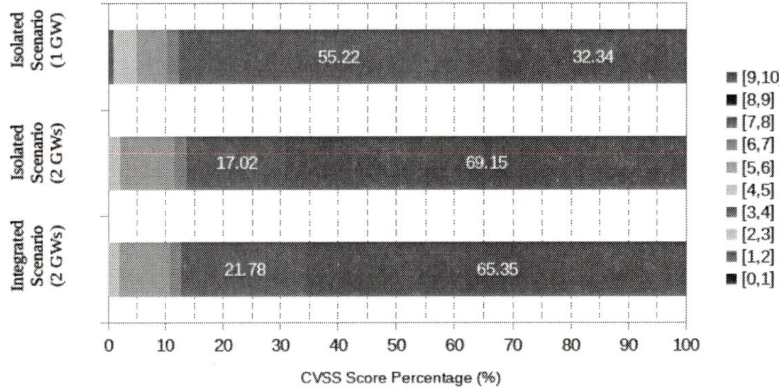

Figure 5.19. *Final comparison of CVSS scores distribution*

The total number of vulnerabilities has naturally increased with the addition of the HA and firewalls as shown in Figure 5.20.

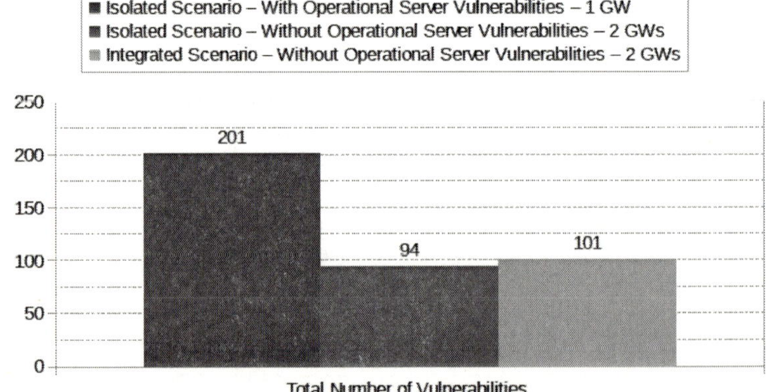

Figure 5.20. *Final comparison of total number of vulnerabilities*

The results comply with the assumptions made in each scenario. However, the most interesting part is probably the comparison between the average CVSS scores that seems to evolve at the opposite of the total number of vulnerabilities: when the total number of vulnerabilities increases, the average CVSS score is likely decreasing (see the results shown in Figure 5.21 for details).

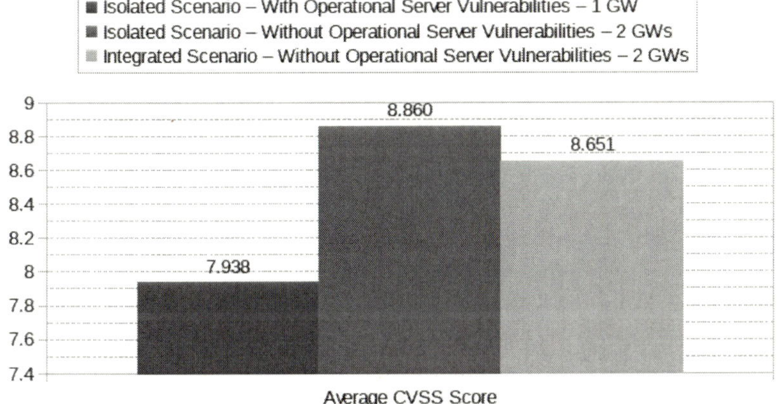

Figure 5.21. *Final comparison of average CVSS score*

This result is important as it helps the security managers to choose the adequate COTS products when it comes to implement their solution and designing the network. In the following sections, the final individual, propagated and network risk results are analyzed and compared to the previous scenarios (with isolated topology).

5.2.7.4. *Individual risk results*

The updated individual risk values relevant to the integrated scenario are presented in Table 5.22.

The nodes in the AeroMACS access service network have not been impacted by the addition of the new nodes, except the ASN gateways protected by a firewall at the entry of the subnetwork. However, all the nodes in the AeroMACS CSN have either the number of security protections increased (because of the addition of the firewalls) or the number of connected nodes decreased. Firewalls do not have the same individual risk values because they are not connected to the same nodes (depending on where there are deployed and what node they protect).

Node ID	Individual risk
Base stations (1–6)	0.23
Base stations 7 and 8	0.3
Base station 9	0.38
Aircraft (all)	0
Vehicle (all)	0
ASN gateway 1	0.24
ASN gateway 2	0.18
DHCP server	12.99
AAA server	3.07
ATS server	0
AOC server	0
APC server	0
HA	1.88
Firewall 1	3.9
Firewalls (2–6)	5.85
Firewalls 7 and 8	11.7

Table 5.22. *Updated individual risk values (integrated scenario)*

Nevertheless, the DHCP server remains the most vulnerable node in the network mainly because of its high and critical COTS vulnerabilities. The individual risk values for the ASN gateways have also decreased because of their lower value in this scenario (Indeed, they are connected to a lower number of nodes compared to the previous scenarios). Unlike the propagated or the network risk, it is useless here to compare the values between the different nodes because individual risks characterize the node itself, and are not related to the topological assumptions made in each simulation. However, the individual risks slightly vary, depending on the connected nodes.

The comparison is more relevant to propagated and network risk results where the addition of firewalls is going to reduce the data flow exchanges between connected nodes (and consequently the likelihood of propagation of threats between them). The node connectivity has also been impacted because of the deployment of firewalls at the entry of the AeroMACS access service network: in this way, ASN gateways are connected to a lower number of nodes and should have their propagated risk values decreased.

5.2.7.5. *Propagated risk results*

As expected, adding the firewalls to the integrated AeroMACS topology has the advantage of getting lower propagated risk values for the connected nodes as shown in Table 5.23.

The most important part here is obviously the decrease of the ASN gateway propagated risk values, which are no longer the points of failure of the integrated topology. Indeed, the addition of the firewalls completely affects the connectivity between nodes. On the one hand, firewalls are deployed now between the servers and the ASN gateways, reducing the connectivity from 9 and 10 to, respectively, 5 and 6. On the other hand, data flows are reduced, inducing a lower likelihood of propagated risk between nodes.

5.2.7.6. *Network risk results*

Table 5.24 summarizes the final node risk results for the integrated AerMACS scenario.

As both the individual and propagated risk values decreased, so do the node risk for all nodes. Obviously, the resulting overall network risk is lower in this scenario compared to all the previous scenarios. The network results are clearly better in the integrated scenario as depicted in Figure 5.22.

Node	Propagated risk
Base stations (1–4)	2.429
hline Base stations 5 and 6	2.576
hline Base stations 7 and 8	5.587
hline Base station 9	5.911
hline Aircraft (1–4)	0.093
hline Aircraft 5 and 6	0.095
hline Aircraft (7–12)	0.121
Vehicle (1–6)	0.093
Vehicles 7 and 8	0.121
Vehicles 9 and 10	0.394
ASN gateway 1	87.225
ASN gateway 2	79.452
DHCP server	0.341
AAA server	0.215
ATS server	0.056
AOC server	0.221
APC server	0.008
HA	0.216
Firewall 1	0.015
Firewalls (2–6)	15.120
Firewalls 7 and 8	45.781

Table 5.23. *Updated propagated risk values (integrated scenario)*

Also, the ASN gateway contribution in the network risk is now more balanced than in the previous scenario as shown in Figure 5.23.

In conclusion, this figure summarizes the approach followed in this work and in the simulation campaign: at each step, the results have been enhanced by making adequate assumptions and by learning what should be learnt from the previous simulation results, using the risk methodology as a decision-making tool. It was quite useful to reduce the network risk from 7858.14 (first scenario) to 698.05 (last scenario), showing that the different predictions, guidance and assumptions were completely done rightly.

Node	Propagated risk
Base stations (1–4)	2.429
Base stations 5 and 6	2.576
Base stations 7 and 8	5.587
Base station 9	5.911
Aircraft (1–4)	0.093
Aircraft 5 and 6	0.095
Aircraft (7–12)	0.121
Vehicle (1–6)	0.093
Vehicles 7 and 8	0.121
Vehicles 9 and 10	0.394
ASN gateway 1	87.225
ASN gateway 2	79.452
DHCP server	0.341
AAA server	0.215
ATS server	0.056
AOC server	0.221
APC server	0.008
HA	0.216
Firewall 1	0.015
Firewalls (2–6)	15.120
Firewalls 7 and 8	44.781

Table 5.24. *Updated network risk values (integrated scenario)*

5.3. Improving AeroMACS network security

Based on the experimental results introduced in the last section several network design guidances can be proposed to be able to improve the global security level of the network architecture. Those guidances are detailed in the following sections.

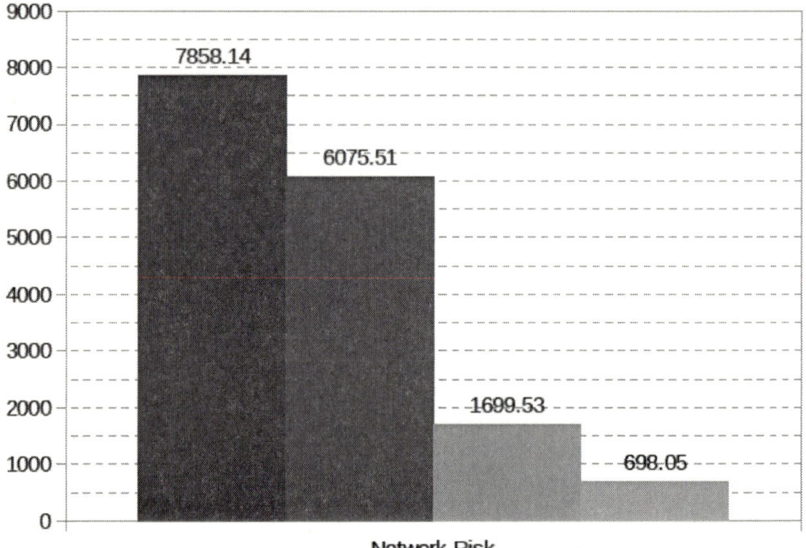

Figure 5.22. *Final comparison of network risk values*

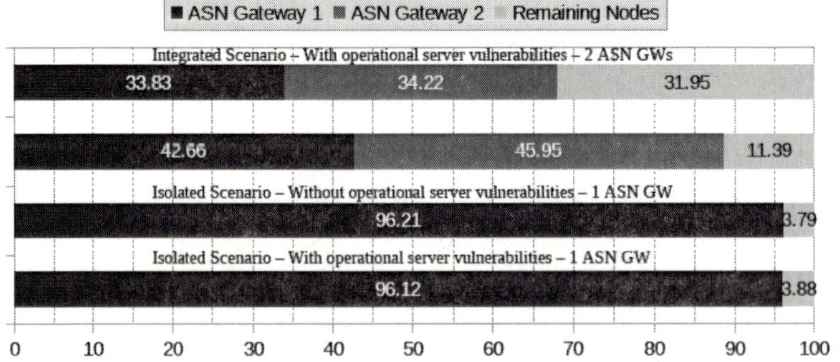

Figure 5.23. *Final comparison of the percentage of network risk per node risk*

First of all, from a security point of view, the central element of an AeroMACS network is probably the AAA server that performs many functionalities beyond its main purposes (meaning authentication, authorization and accounting). Indeed, the AAA server is widely used to supply user-related information, such as QoS parameters or ASN network configuration, but it is also involved in IP mobility procedures (such as handover procedures) or even in IP host configuration. DHCP is also important from a security point of view because some attacks can be conducted, but fortunately can be avoided as depicted in the following sections.

5.3.1. *DHCP security*

In an AeroMACS network, the DHCP protocol can be configured in two ways: using either a DHCP proxy or a DHCP relay. Note that the DHCP protocol is solicited when the first packet sent by the Mobile Station (MS) is a DHCP_DISCOVER frame:

– When using a DHCP proxy, the ASN answers to the DHCP_REQUEST frame sent by the MS and assigns it the IP address. In most cases, if a seamless handover is required, meaning that the session has to be kept alive, the assigned IP address should be the same before and after the handover. To do this, the MS puts the IP address he was already using in a specific option of the DHCP_DISCOVER frame. However, the DHCP_DISCOVER frames are not signed (i.e. unauthenticated) which means that a malicious MS could be trying to obtain the IP address of a legitimate MS. To avoid this security issue, the AAA server is involved in the IP address assignment procedure, meaning that the effective IP address to be assigned to the MS will be included into a specific Remote Authentication Dial In User Service (RADIUS) attribute, part of the ACCESS_ACCEPT packet coming from the AAA server in the CSN.

– When using a DHCP relay, the ASN does not answer to the MS but forwards the request to a remote server. If the DHCP relay does not know the IP address of the remote server, this information has to be included in a RADIUS attribute that is received by the ASN at the end of the authentication (note that the remote server must be the same server used by the MS to be authenticated for the first time). Using a DHCP relay to communicate with a remote DHCP server is a tricky procedure as long as both integrity and packet authentication are not provided, leading to several attacks, such as DNS spoofing (this can be done by modifying the DNS server address in order to redirect MSs to fake URLs) or man-in-the-middle attacks by changing the default gateway address. To address this security issue, the DHCP protocol introduces a security extension [DRO 01] to authenticate the frames exchanged between the DHCP relay agent and the DHCP server using a pre-configured symmetric key between them. This key, called DHCP_RK, is used to derive other keys to secure each single session. To dynamically redirect the DHCP_REQUEST frames, the DHCP_RK must also be dynamically moved into the DHCP relay and the DHCP server when need. To do so, the AAA server is once again involved using RADIUS attributes. Figure 5.24 shows the DHCP protocol key exchange when a DHCP relay is used.

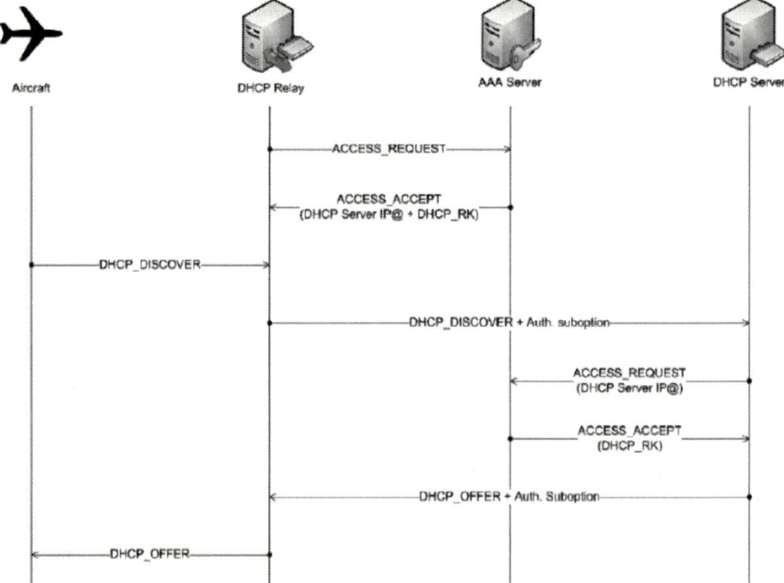

Figure 5.24. *DHCP key management using a DHCP relay*

The exchange scheme shown in Figure 5.24 works as follows:

– When an MS is being authenticated, the CSN generates a random key and assigns it to the AAA server.

– The DHCP relay sends an ACCESS_REQUEST to the AAA server, which responds using an ACCESS_ACCEPT packet containing the DHCP server IP in the CSN, the DHCP_RK key, its lifetime and a unique ID, all encapsulated into a RADIUS attribute.

– When the ACCESS_ACCEPT packet is received by the DHCP relay, it uses it to sign the DHCP_DISCOVER frame sent to the DHCP server in the CSN.

– If the DHCP server does not already know the DHCP_RK key, it asks the AAA server, otherwise it uses it to verify the signature of the DHCP_DISCOVER frame sent by the DHCP relay.

– If the signature verification is correct, the DHCP server sends back a DHCP_OFFER packet, which will be forwarded by the DHCP relay to the MS.

Note that a RADIUS client has to be implemented on both DHCP relay and server in order to communicate correctly with the AAA server (RADIUS operates in a pure client-server paradigm, meaning the server does not initiate any messages but only replies to client requests).

5.3.2. *Mobile IP security*

AeroMACS uses the Mobile IP (MIP) protocol [PER 02], [CAL 07] to provide seamless, transparent and flexible IP addressing in the network. The main intent is to allow an MS to have a permanent IP address while moving from one point of access to another, especially in roaming scenarios. Using DHCP allows the MS to keep a session alive and to have the same IP address before and after a handover, but this only works in one direction, meaning that the other communication party will receive IP packets but its responses will be routed to the home network of the MS. The MIP protocol is then used to forward these packets from the home network to the visited network. When using the MIP protocol, an MS has two IP addresses:

– A home address (also called home of address HOA) that is the IP address assigned by the HA in the home network. This address remains fixed while the MS is roaming and is always used as long as the MS remains under its home network coverage.

– A care of address, (COA) given to the MS by a foreign agent (FA) in the visited network. After receiving the COA address, the MS sends a REGISTRATION_REQUEST (Mobile IP RRQ) to its HA that contains its COA address, which is needed to reach the MS in the visited network. The HA confirms the reception by sending back a REGISTRATION_REPLY (MIP RRP). When the MS is communicating from the visited network with a correspondent end entity, it uses a direct route from the visited network to reach the destination. However, when the correspondent node wants to communicate with that MS, it has to send the IP packets using the MS home address. When the HA receives these packets, it looks to the associated COA address, sends the packets to the adequate FA, which finally forwards the packets to the MS.

Security issues related to MIP depends on how MIP is supported on the MS side. There are two kinds of MIP support on the MS:

1) Proxy MIP (PMIP): this is used for MSs that are not MIP compliant but still need persistent connections and seamless roaming when moving from a home network to a visited network. In this case, a third-party entity called PIMP Mobility Manager is in charge of handling the MIP registration procedure for the MS toward the FA by intercepting the DHCP_DISCOVER frame sent by the MS (in order to get the COA address in the visited network), and performs the MIP registration with the HA. When the registration is finished, the PMIP Mobility Manager uses the DHCP protocol to assign the COA IP address to the MS. Again, the AAA server is involved in this procedure as the PMIP Mobility Manager receives multiple information (such as the MS HOA or the HA IP address) encapsulated in specific RADIUS attributes. Figure 5.25 shows all the logical entities involved when an MS is PMIP compliant.

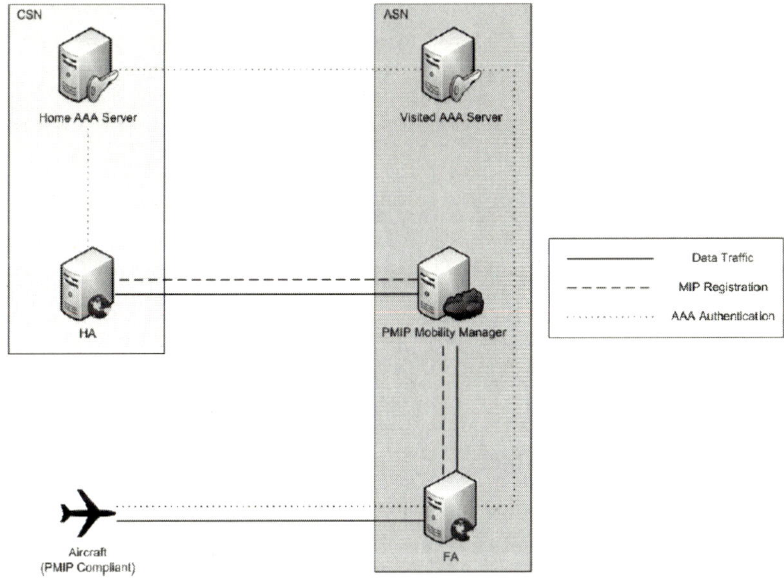

Figure 5.25. *Mobile IP registration – PMIP case*

When an MS receives beaconing traffic from a foreign network it wants to access, it starts an authentication process with the H-NSP as depicted in Figure 5.26:

– The authenticator (which is co-located with the PMIP Mobility Manager) intercepts the ACCESS_REQUEST sent by the MS and forwards it to the Home AAA server.

– The AAA server sends back an ACCESS_ACCEPT packet that contains information needed by the authenticator to contact the HA (i.e. IP address) and encrypts the MIP registration messages (i.e. the Master Session or MSK key produced by the AAA server).

– The authenticator communicates the MSK key to the PMIP Mobility Manager, which uses it later to derive the MIP_RK and the HA_RK keys needed to generate respectively the MS_HA key (to secure the MIP_RRQ message) and the FA_HA key (to secure the messages between the FA and the HA agents). Note that there is no need in this case for the MS_FA key because the MS and the FA are located in the same ASN.

– The PMIP Mobility Manager is now able to proceed to an MIP registration on behalf of the MS by sending a signed MIP_RRQ packet to the HA using the MN_HA key.

– To verify the MIP_RRQ packet signature and continue the Mobile IP registration process, the HA sends an ACCESS_REQUEST packet to the AAA server asking for the needed keys (i.e. MN_HA and HA_RK keys).

– The AAA server responds with an ACCESS_ACCEPT RADIUS packet that contains the required keys for the signature verification.

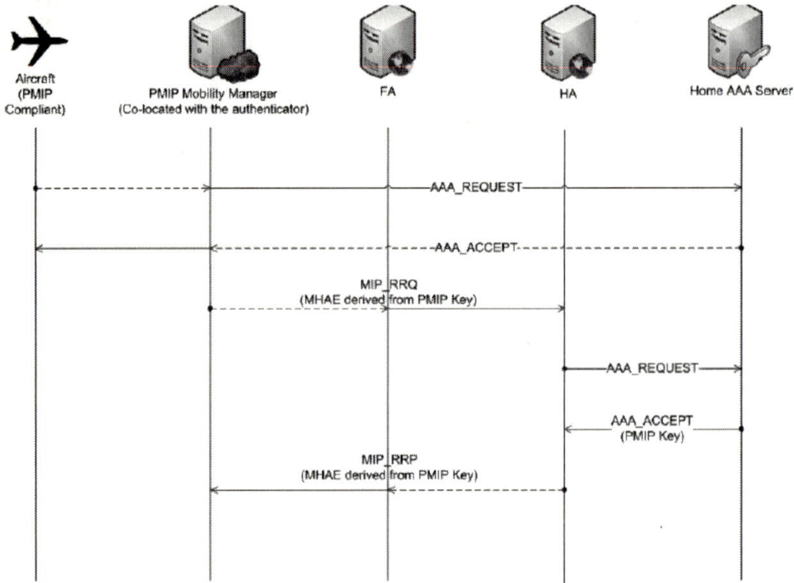

Figure 5.26. *Mobile IP key management – PMIP case*

2) Client MIP (CMIP): this is used for MSs that completely support MIP. Figure 5.27 shows the MIP registration for CMIP compliant MSs, which does not need a mobility manager in this case. Moreover, the FA is able to distinguish between PMIP and CMIP clients depending on the first packet they send respectively (PMIP clients send a DHCP_DISCOVER packet whereas CMIP clients send an MIP_RRQ packet instead).

The security procedure is practically the same as for the PMIP clients except that the usage of the FA_RK key is needed to derive the MS_FA key (which is used to provide integrity and message authentication between the MS and the FA).

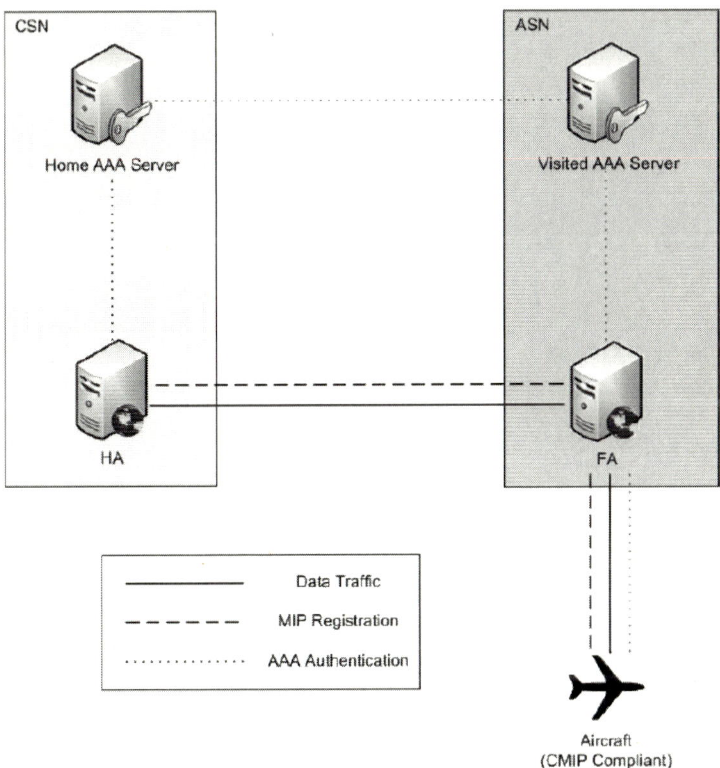

Figure 5.27. *Mobile IP registration – CMIP case*

Conclusion

In this book, a new quantitative risk assessment method for network security based on risk propagation has been described. The proposed methodology estimates the network risk level quantitatively based on several criteria such as the complexity of the conducted attack, or its impact within the network. Vulnerability statistics issued from the NVD public database are used within the methodology throughout the CVSS impact scores.

The main model components have been defined and their consistency has been detailed. The drawbacks of existing methods have been discussed and then the identified theoretical drawbacks have been palliated. Part 1 of this book has been dedicated to quantifying the main risk parameters used in the risk assessment methodology. Part 2 has dealt with experimental results relative to an aeronautical case study in the scope of the SESAR project. The main goal was to conduct a risk assessment of network security for the AeroMACS technology intended to be deployed in future airport surface area communications.

The AeroMACS security has been analyzed using the risk assessment methodology; the experimental results highlighted several weaknesses of the AeroMACS system in an isolated network topology (i.e. without additional security features), meaning that the system needs more consideration from a security point of view. Following several areas of guidance drawn after the preliminary results, the AeroMACS network topology has been improved step-by-step in order to reduce the network security risk. Comparison between all experimental scenario results illustrates that this new risk assessment methodology can be used as a decision-making tool in network and security administration.

Bibliography

[ABO 04] ABOBA B., BLUNK L., VOLLBRECHT J., CARLSON J., LEVKOWETZ H., Extensible authentication protocol (EAP), RFC 3748 (Proposed Standard), June 2004. Updated by RFC 5247.

[AHM 08] AHMED M.S., AL-SHAER E., KHAN L., "A novel quantitative approach for measuring network security", *Proceedings of INFOCOM 2008. The 27th Conference on Computer Communications, IEEE*, pp. 1957–1965, 2008.

[BAR 05] BARBEAU M., "Wimax/802.16 threat analysis", *Proceedings of the 1st ACM International Workshop on Quality of Service & Security in Wireless and Mobile Networks*, Q2SWinet '05, ACM, New York, NY, pp. 8–15, 2005.

[BAR 09] BARTH A., RUBINSTEIN B.I.P., SUNDARARAJAN M., MITCHELL J.C., SONG D.X., BARTLETT P.L., "A learning-based approach to reactive security", *CoRR*, abs/0912.1155, 2009.

[BEN 92] BENNETT S.P., KAILAY M.P., "An application of qualitative risk analysis to computer security for the commercial sector", *Proceedings of the Eighth Annual Computer Security Applications Conference*, pp. 64–73, 1992.

[CAL 07] CALHOUN P., PERKINS C., BHARATIA J., "Mobile ipv4 challenge/response extensions (revised)", *IETF*, 4721, January 2007.

[CHE 06] CHEN X., ZHENG Q., "Quantitative hierarchical threat evaluation model for network security", *Journal of Software*, vol. 17, no. 4, pp. 885–897, April 2006.

[DAN 09] DANFENG Y., FANGCHUN Y., YU L., "Service-based quantitative calculation of risk for ngn", *Proceedings of the 2nd IEEE International Conference on Broadband Network and Multimedia Technology, IC-BNMT '09*, pp. 306–310, 2009.

[DCS 04] DCSSI, Ebios: Introduction, démarche, techniques, outillage, 2004.

[DRO 01] DROMS R., ARBAUGH W., "Authentication for dhcp messages", *IETF*, vol. 3118, June 2001.

[ERN 08] EREN E., DETKEN K.-O., "Wimax security – assessment of the security mechanisms in IEEE 8o2.16d/e", *The 12th World Multi-Conference on Systemics, Cybernetics and Informatics: WMSCI 2008*, Orlando, FL,2008.

[ETS 03] ETSI, Telecommunications and internet protocol harmonization over networks (tiphon) release 4; protocol framework definition; methods and protocols for security; part 1: threat analysis, 2003.

[EUR 02] EUROCONTROL, Communications operating concept and requirements for the future radio system, 2002.

[EUR 11] European Commission EUROCONTROL, First sesar release 2011, 2011.

[FAA 11] FAA, Faa's nextgen implementation plan, March 2011.

[FRE 02] FREDRIKSEN R., KRISTIANSEN M., GRAN B., STØLEN K., OPPERUD T., DIMITRAKOS T., "The CORAS framework for a model-based risk management process", in ANDERSON S., FELICI M., BOLOGNA S., (eds), *Computer Safety, Reliability and Security*, Chapter 11, Lecture Notes in Computer Science, vol. 2434, Springer Berlin/Heidelberg, Berlin, Heidelberg, pp. 39-53, September 2002.

[FRI 08] FRIGAULT M., WANG L., "Measuring network security using bayesian network-based attack graphs", *Proceedings of the 32nd Annual IEEE International Computer Software and Applications, COMPSAC '08*, pp. 698–703, 2008.

[HAL 10] HALL E., BUDINGER J., DIMOND R., WILSON J., APAZA R., "Aeronautical mobile airport communications system development status", *Proceedings of the Integrated Communications Navigation and Surveillance Conference (ICNS)*, 2010.

[HUA 09] HUANG R., YAN D., YANG F., "Research of security metric architecture for next generation network", *IEEE International Conference on Network Infrastructure and Digital Content 2009, IC-NIDC 2009*, pp. 207–212, November 2009.

[ICA 05] ICAO, ICAO manual on required communications performance, appendix N (manual on RCP) to the report, Technical Report Version 5.1, September 2005.

[ISO 99] ISO, ISO/IEC 15408: common criteria for information technology security evaluation, 1999.

[ISO 04] ISO, Information technology - security techniques – management of information and communications technology security – part 1: concepts and models for information and communications technology security management, 2004.

[ISO 05a] ISO, ISO/IEC 27001: information technology, security techniques, information security management systems requirements, 2005.

[ISO 05b] ISO, ISO/IEC 27002: information technology – security techniques – code of practice for information security management, 2005.

[ISO 08] ISO, ISO/IEC 27005: information technology, information security techniques, information security risk management, 2008.

[JIN 09] JING L., "Risk evaluation process model of information security", *Proceedings of the International Conference on Measuring Technology and Mechatronics Automation, ICMTMA '09*, vol. 2, pp. 321–324, 2009.

[JOH 04] JOHNSTON D., WALKER J., "Overview of IEEE 802.16 security", *Security and Privacy*, vol. 2, no. 3, pp. 40–48, May 2004.

[KON 10] KONDAKCI S., "A causal model for information security risk assessment", *Proceedings of the 6th International Conference on Information Assurance and Security (IAS) Conference*, pp. 143–148, 2010.

[KRA 97] KRAWCZYK H., BELLARE M., CANETTI R., HMAC: Keyed-Hashing for Message Authentication, RFC 2104 (Informational), February 1997.

[LAO 08] LAO G., WANG L., "The quantification management of information security risk", *Proceedings of the 4th International Conference on Wireless Communications, Networking and Mobile Computing, WiCOM '08*, pp. 1–4, 2008.

[LOD 08] LODDO J.V., SAIU L., "Marionnet: a virtual network laboratory and simulation tool", *Simulation Works*, 2008.

[LV 09] LV H., "Research on network risk assessment based on attack probability", *Proceedings of the 2nd International Workshop on Computer Science and Engineering, WCSE '09*, vol. 2, pp. 376–381, 2009.

[MEN 09] MENG Q., DONG M.H., LI Y., MING Z.W., "Network security analysis model based on logic exploitation graph", *Computer Engineering*, vol. 9, pp. 147–149, 2009.

[MIC 04] MICROSOFT, *The Security Risk Management Guide*, 2004.

[NAS 09] NASA, Initial assessment of security provisions of the IEEE 802.16e-2005 standard for airport surface communications, Technical report, 2009.

[NIS 02] NIST, Federal information processing standards publication (fips-pubs) 800-30, risk management guide for information technology systems, July 2002.

[NIS 05] NIST, Recommendation for block cipher modes of operation: the cmac mode for authentication, NIST special publication 800-38b, 2005.

[PER 02] PERKINS C., "Ip mobility support for IPV4", *IETF*, vol. 3344, August 2002.

[SCH 05] SCHIFFMAN M., A complete guide to the common vulnerability scoring system (CVSS), June 2005.

[SHE 06] SHER M., MAGEDANZ T., "Developing network domain security (NDS) model for IP multimedia subsystem (IMS)", *Journal of Networks*, vol. 1, no. 6, 2006.

[WIL 11] WILSON S., "The network security architecture and possible safety benefits of the aeromacs network", *Proceedings of 2011 Integrated Communications Navigation and Surveillance (ICNS) Conference*, 2011.

[WIN 08] WING J.M., "Scenario graphs applied to network security", *Information Assurance*, Morgan Kaufmann, Burlington, MA 2008.

[XU 06] XU S., HUANG C.-T., "Attacks on pkm protocols of IEEE 802.16 and its later versions, 2006", *3rd International Symposium on Wireless Communication Systems, 2006, ISWCS '06*, 2006.

[YAU 99] YAU S.S., ZHANG X., "Computer network intrusion detection, assessment and prevention based on security dependency relation", *Proceedings of the 23rd Annual International Computer Software and Applications Conference, COMPSAC '99*, pp. 86–91, 1999.

[YLO 06] YLONEN T., LONVICK C., The secure shell (SSH) protocol architecture, RFC 4251 (Proposed Standard), January 2006.

[YON 08] YONGLI T., GUOAI X., YIXIAN Y., "Information security management measurement model based on AHP", *Journal of Liaoning Technical University*, vol. 27, 2008.

[ZHA 04] ZHANG Y.-Z., FANG B.-X., YUN X.-C., "A risk assessment approach for network information system", *Proceedings of 2004 International Conference on Machine Learning and Cybernetics, 2004*, vol. 5, pp. 2949–2952, August 2004.

Index

O

P, Q

R

S

T, U, V

W